W0037510

Out of Nazi Germany in Time, a Gift to American Science

Gerhard Schmidt, Biochemist

Gerhard Schmidt.

Out of Nazi Germany in Time, a Gift to American Science

Gerhard Schmidt, Biochemist

B. David Stollar

American Philosophical Society Press
Philadelphia • 2014

Transactions of the
American Philosophical Society
Held at Philadelphia
For Promoting Useful Knowledge
Volume 104, Part 1

Copyright © 2014 by the American Philosophical Society for its Transactions series.

All rights reserved.

ISBN: 978-1-60618-041-9
US ISSN: 0065-9746

Library of Congress Cataloging-in-Publication Data

Stollar, B. David, 1936- , author, editor.
   Out of Nazi Germany in time, a gift to American science : Gerhard Schmidt,
biochemist / B. David Stollar.
      p. ; cm. — (Transactions of the American Philosophical Society, 0065-9746 ; v. 104,
pt. 1)
   Includes bibliographical references and index.
   ISBN 978-1-60618-041-9 (alk. paper)
   I. American Philosophical Society, publisher. II. Title. III. Series: Transactions of the
American Philosophical Society ; v. 104, pt. 1. 0065-9746
      [DNLM: 1. Schmidt, Gerhard, 1901-1981. 2. Biochemistry—history—Germany—
Biography. 3. Holocaust—history—Germany—Biography. 4. National Socialism—history—
Germany—Biography. 5. World War II—Germany—Biography. WZ 100]
   QP511.8.S35
   572.092—dc23
   [B]
                              2014039641

The author and publisher are grateful to the family of Gerhard Schmidt for graciously giving
permission to print photographs from family albums.

Dedicated to the memory of those who did not escape.

Contents

Foreword

David Stollar, in his memoir of Professor Gerhard Schmidt, has accomplished a remarkable literary feat. Embedded in this elegant study of an important scientific career are several narratives that are invaluable for anyone interested in the forces that shaped both science and history in the first half of the twentieth century. Most significant for both an American and German audience is the story of the destruction of the German university system and German science at the hands of Adolf Hitler and Nazism. There is drama, tragedy, heroism, and just plain good fortune for some in this often frightening adventure. Along the way, Professor Stollar takes us on a journey of German intellectual and social history, from pre-World War I Wilhelminean prosperity that chronicles the integration of German Jews into a society in which they thought they were welcomed; to the rollercoaster postwar world of Weimar culture, and the looming threat of German anti-Semitism, even before the coming of Hitler in 1933. Then, suddenly, hundreds of German professors, physicians, and researchers were dismissed from university positions, and within a few years most were refugees. Those who felt most German were stunned into disbelief and despair. Some, like Gerhard Schmidt, had the will and tenacity to survive; others less fortunate ended their lives at their own hands or in German concentration camps.

That is the terrible part. The happy story is that many were saved, thanks to the efforts of fellow scientists living in freedom, as well as the aid of some extraordinary organizations and universities. Out of the misery caused by dislocation and the tragedies of lives disrupted and even lost, came the remarkable intellectual explosion that placed the United States at the pinnacle of higher education and scientific research. Adolf Hitler provided American faculties—up until the 1930s for the most part, and with a few notable exceptions, undistinguished colleges and universities with modest research agendas—with the means to become the envy of the world; and we did not miss the opportunity. America opened up its otherwise shut immigration doors just a crack to allow in German scientists who helped establish American research hegemony for decades to come. They gave their Nobel Prize speeches in heavily accented English, and as American citizens. It is arguable that, had Gerhard Schmidt enjoyed a more orderly life in exile, he would have been one of those Nobel laureates.

There are stories inside stories. For my institution, Tufts University, Professor Stollar points to the remarkable cooperation between two physicians—one Jew, one Gentile—in providing a haven for refugee Germans. Sam Proger and Joseph Pratt, along with Tufts President Leonard Carmichael—reached out, and in doing so gave Tufts both moral and intellectual grounds to be proud.

Out of the trials and tribulations of a world gone mad, Gerhard Schmidt emerges as an individual who tenaciously held on to make a life for himself and his family in a new country. This is an exciting story, well told by a colleague with the gift of historical memory.

Sol Gittleman
Alice and Nathan Gantcher University Professor
Tufts University

Acknowledgments

This work depended on the vision of Drs. Gerhard Schmidt and Morris Cynkin, who recorded conversations they shared in the early 1970s, with a goal of preserving an important story. Thanks are due to Drs. Henry Mautner and Clark Sawin, who arranged the partial transcription of the recorded discussions. I give special thanks to members of Gerhard's family—Mrs. Edith Schmidt, Dr. Milton and Homai Schmidt, Michael Schmidt, and Dr. Renate Bever—who have generously shared their memories and access to family history and provided numerous photographs. Recollections of working with Gerhard were shared by his students: Drs. Lowell Greenbaum, Maurice Bessman, Marya Seraydarian, and Peter Cashion.

Many people have provided sources of information. I thank Anne Sauer, Director of Tufts University's Digital Collections and Archives, where the cassettes with the recorded conversations are deposited; Dr. Susanne Belovari, PhD Archivist for Reference and Collections at Tufts University, for help on records of Gerhard and other German refugee physicians at Tufts University School of Medicine; and Elizabeth Richardson, Associate Librarian of Tufts Health Sciences Library for help in the School of Medicine archives. Dr. Alan F. Hofmann provided a copy of his book, *Siegfried Thannhauser (1885–1962) Physician and Scientist in Turbulent Times*. Dr. David Zimmerman, Department of History, University of Victoria in Canada, published important insight related to Dr. Schmidt's two-year appointment at Queen's University, and provided copies of documents from the files of the Society for the Protection of Science and Learning (now the Council for Assisting Refugee Academics [CARA]; I thank Ryan Mundy of CARA for permission to cite these documents. Susan Office, Assistant to the University Archivist of Queen's University, also provided documents related to the years in Canada. I am grateful to Prof. Ulrich Brandt, Department of Molecular Bioenergetics of the Gustav Embden Center for Biological Chemistry, Johann Wolfgang Goethe-Universität in Frankfurt, Germany, for providing a biography of Dr. Gustav Emden, Gerhard's mentor, in the form of a 1992 doctoral thesis by Dr. Ulrich Flaig. Additional information on Dr. Embden was provided by Prof. Lothar Jaenicke of the Institute for Biochemistry, Universität zu Köln. Laura Slezak Karas, Manuscripts Specialist at the New York Public Library, sought records in archives of the Emergency Committee in Aid of Displaced Foreign Scholars. Dr. Stephen Soltoff helped retrieve early Schmidt publications. Daniel Bird, Director of Volunteer Services at the Tufts Medical Center, provided a copy of Herbert Black's book, *Doctor and Teacher, Hospital Chief*, on Dr. Samuel Proger and the history of the Boston Dispensary/Boston Floating Hospital/Tufts University collaboration in the New England Medical Center. Dr. Rosemary Polsky-Newman provided a photo of Dr. Morris Cynkin, her late husband.

Evaluation and feedback are essential to development of such a project. I am grateful to Lawrence Stollar, Carol Stollar, and Dr. Michael Nevins for reading the manuscript and providing valuable suggestions; and to Professor Sol Gittleman for both reading the manuscript and writing a Foreword and for sharing interest in and beyond this project. I also thank Professor Brian Schaffhausen and the Department of Biochemistry at Tufts for the opportunity to present part of this project as the 2012 Gerhard Schmidt Memorial Lecture. I am also grateful to Miriam Spectre for encouraging me to contact the American Philosophical Society, and to Mary McDonald, Director of Publications for the Society, for critical guidance in the many steps involved in the process of publication. I add special thanks to Pamela Lankas for critically important and expert editorial help and guidance in preparation for publication.

I am deeply grateful to my wife, Carol, for loving support and patience as well as feedback during this project.

Introduction

In March 1973, a symposium was held at Tufts University to honor Gerhard Schmidt, a biochemist who had recently been elected to membership in the United States National Academy of Sciences. Participating speakers included leaders in the field of biochemistry: Erwin Chargaff, David Nachmanson, and Nobel Prize recipients Fritz Lipmann, Carl Cori, and Severo Ochoa. Other similarly distinguished guests were in the audience. They were friends of Gerhard, and, like him, many were key figures in the history of biochemistry and in the story of Europe's contribution to American science. One of those friends, Herman Kalckar, wrote a biographical essay about him for the National Academy of Sciences.[1]

We have a way of learning a good deal more about Gerhard Schmidt and the context of his life and work because, between 1971 and 1973, he sat together, every two or three weeks, with Morris Cynkin,[2] a colleague in the Tufts University School of Medicine's Department of Biochemistry, and they recorded conversations about Gerhard's story. Cynkin introduced the interview sessions, stating: "When I first came to Tufts Medical School in 1960 as a young assistant professor, one of my senior colleagues was Professor Gerhard Schmidt. Before I came to Tufts, Schmidt was a name that I had come to know with regard particularly to a procedure known as the Schmidt–Thannhauser procedure, which was used for the preparation of nucleic acids. In the years

---

[1] Kalckar, H., "Gerhard Schmidt 1901–1981," *Biographical Memoirs of the National Academy of Sciences*, 57 (1987): 399–429.

[2] Morris Cynkin (1930–87), a native of Brooklyn, New York, earned a Bachelor of Science degree from City College of New York in 1952, and, studying with Martin Gibbs, a PhD in Bacteriology from Cornell University, in 1956. He continued in postdoctoral work with Gibbs for two years and was then a Visiting Scientist at the National Institutes of Health, in the Laboratory of Biochemistry and Metabolism, in Bethesda, MD. In 1960 he joined the faculty of the Department of Biochemistry at Tufts University School of Medicine, where he rose to the rank of professor in 1971 and continued in research and teaching through the rest of his life. His research career, launched with ten years of support through a Career Development Award from the National Institutes of Health, concerned carbohydrate metabolism and the structure and biosynthesis of polysaccharides and glycoproteins. He had a strong interest in the history of discovery in biochemistry and other fields of science.

since then, I've come to know Gerhard Schmidt as a friend and a senior colleague and he has given me great cause for affection and admiration. He is a difficult man at times and stubborn. He's also cultured, compassionate, wise, and very often amusing. About two years ago, the thought came to me that it really is a shame that of all the amusing and enlightening stories that Gerhard has told me over the years, none of them has been recorded in such a way that other people could share the experience that I have had in my many conversations with Gerhard Schmidt. The interviews that follow this statement are my attempt to share with colleagues and students and any other people who are interested in human beings, the experiences of a very interesting scientist and human being."

Drs. Schmidt and Cynkin left 30 cassette tapes, with about 40 hours of recorded discussions. Henry Mautner,[3] then Chairman of the Department of Biochemistry and Pharmacology at Tufts (who also organized the 1973 symposium honoring Schmidt), arranged the beginning of the transcription of the interview tapes. After Mautner's passing in 1995, Clark Sawin, an endocrinologist and educator at Tufts University School of Medicine, who was also active in studies of the history of medicine, arranged further transcription; but much of the discussion remained only on the tapes when Sawin moved from Tufts to the Veterans Administration in Washington, DC, in 1998. As I was Chairman of the Department of Biochemistry at that time, Dr. Sawin returned the tapes to me in January 2001.

Having known Gerhard Schmidt personally since joining the Medical School faculty at Tufts in 1964, I was interested in the project and had followed the progress of preserving these conversations. In fact, I had benefitted from Dr. Schmidt's generosity before ever meeting him. As a postdoctoral fellow at Brandeis University from 1960–62, I had a need for a specific enzyme, a *phosphatase* that would remove the phosphate ester group at the ends of nucleotide chains (DNA fragments) without cleaving within the chains. Someone suggested that Professor Schmidt at Tufts might provide such an enzyme. He did so, sending much more than enough for the completion of that project.

Converting the audio content of the interview tapes to computer sound files, I completed the transcription, edited and annotated the transcripts, and

---

[3] Henry Mautner (1925–95), born in Prague, Czechoslovakia, earned a bachelor's degree in chemistry from the University of California at Los Angeles in 1946, a master's degree from the University of Southern California in 1949, and his PhD from the University of California at Berkeley in 1955. He served on the faculties of the University of Southern California (1947–49), University of California (1951–55), and Yale University (1956 to 1970), where he became a professor and head of Medicinal Chemistry. In 1970, he moved to Tufts University School of Medicine to chair a newly combined Department of Biochemistry and Pharmacology. He was known for research on organic compounds containing selenium and their anticancer potential, as well as for work on acetylcholine receptors. He had been actively engaged in the Children's School of Science in Woods Hole, MA, and, after retirement from the Tufts faculty in 1984, served as science chairman for that school; Banks, H. H. *A Century of Excellence: The History of Tufts University School of Medicine 1893–1993* (Boston: Tufts University Press, 1993; *New York Times* obituary April 15, 1995.

explored questions arising out of their content.[4] These interviews opened doors to—and led me to study more about—several realms of history: Gerhard's personal and scientific story, pioneers and foundational research in biochemistry, Jewish life in Weimar Germany, the transfer of outstanding scientists from Europe to America, great institutions in which Gerhard worked during the seven-year odyssey on his way to Boston, and Tufts University School of Medicine.

The narrative is based fundamentally on the recorded interviews, bringing into a rough chronological order some discussions that were not sequential in the transcripts and tying together some fragmented conversations. Some comments and descriptions of events are identified as direct quotations. Edith (Straus Horkheimer) Schmidt's writings about major experiences in her life shed light on personal history as well. Additional information has been added, on historical events in Germany and Italy, for example, related to the events that Gerhard recalled during the conversations. Footnotes provide specific references to scientific findings and events mentioned in the recordings. Some additional scientific information has been added—based, for example, on the content of Gerhard's publications, as well as biographical information on scientists with whom he worked. Discussions on the tapes cover events and research up to the early 1960s. Material on later research comes from study of publications and interviews.

Some additional information comes from personal knowledge. Once I joined the Tufts faculty, Gerhard and I were colleagues for 17 years, until his death in 1981, and we met in various settings: lecture halls, research seminars, and department meetings. At first our laboratories were located in different buildings; but in the early 1970s he moved from one of the hospital buildings into a laboratory in our recently formed Department of Biochemistry and Pharmacology, which was in the medical school building, so we met more regularly, sometimes outside the laboratory or department setting. As a personal friend, for example, he kindly invited me to be his guest at a dinner meeting of the American Academy of Sciences, of which he was an elected member.

This project was also aided greatly by conversations with members of Gerhard's family. During his life, Gerhard and his wife, Edith, were warm hosts, inviting me and my wife into their home. Knowing their cultured background and interests has helped in understanding the content of the Schmidt–Cynkin discussions. After Gerhard's passing, it was a privilege to continue my relationship with Edith, a vital, insightful and delightful person throughout her long life, until her death in August 2012, two months before

---

[4]The cassette tapes and digital files of the edited transcripts have been deposited in the archives of Tufts University.

her 102nd birthday. Getting to know their sons, Michael and Milton, and Milton's wife, Homai, and their children, Eric and Alice, was also very important, as they showed great interest in this project and read and responded to portions of the narrative, as did Gerhard's youngest sister, Dr. Renate (Schmidt) Bever. Milton and Homai provided additional insight by sharing treasured albums containing a photographic family record.

One

Origins of a Universalist

What was it that sparked a passion for science in Gerhard Schmidt, a teenage boy growing up in a cultured Jewish home in Stuttgart, Germany, in the early years of the 20th century? Perhaps it was the example of his father, a professor of chemistry; but primarily, Gerhard said, it was exposure to particular books. One of those books, which he received as a gift for his fifteenth birthday, left a profound and lasting impression; it was entitled *Der Mensch (Man)*, and "In this book, there was a color plate depicting fertilization and cell division and a schematic illustration of mitosis and chromosomes, and I had never heard about this and these structures fascinated me tremendously. . . . Of all the things, nothing impressed me more than the picture I saw of chromosomes."[1] Perhaps here was laid the foundation for Gerhard's long-standing interest in nucleic acids and nucleoproteins during his research career.

Gerhard recalled the high quality of popular science publications in Germany at that time, up to and during the years of World War I. He was especially impressed by books of the supporters of Darwinian evolution, led by Ernest Haeckel, professor of zoology in Jena. To popularize science, Haeckel produced *Kunstformen in Natur (Artforms in Nature)*, with wonder-producing illustrations, such as figures of diatoms and Radularia skeletons.[2] The biologist Wilhelm Bolsche wrote many popular articles, a comprehensive history of the Earth, and even a three-volume work: *Love Life in Nature*.[3] Among the articles Bolsche composed was one on "radiation pressure," a concept developed by the British physicist James Clerk Maxwell and described by the Swedish scientist Svante Arrhenius in his book, *The Life of the Universe, As Conceived by Man from the Earliest Ages to the Present Time*.[4] Arrhenius

---

[1] Probably referring to Johannes Ranke, *Der Mensch* (Leipzig: Verlag des Bibliographischen Instituts,1886), a richly illustrated book of anatomy, embryology, and physiology. The 1886 edition included illustrations of cell division, but not a plate of chromosomes. Chromosomes had been described by Walther Flemming and represented in many illustrations in his book *Zellsubstanz, Kern und Zelltheilung* (Verlag von F. C. W. Vogel, 1882). They may have been incorporated into later modified editions of *Der Mensch*.

[2] Ernst H. P. A. Haeckel, *Kunstformen der Natur* (Leipzig: Verlag des Bibliographischen Instituts, 1904). Reprinted versions of the plates are available in English [E.H.P.A. Haeckel, O. Breidbach, and I. Eibl-Eibesfeldt, *Artforms in Nature* (New York: Prestel Publishing, 1904)] and on many sites online.

[3] Wilhelm Bolsche, *Das Liebesleben in der Natur* (Jena: Eugen Diederichs Verlag, 1910).

[4] Svante Arrhenius, *The Life of the Universe, As Conceived by Man from the Earliest Ages to the Present Time*, trans. H.Borns (London: Harper & Bros, 1909), 168.

1

described radiation pressure, the pressure exerted on any surface exposed to electromagnetic radiation, as a force by which material might be transported from star to star. Bolsche's article ended with the dramatic sentence: "Perhaps we can imagine that a hundred million years from now, driven by Arrhenius' radiation pressure, a little dust particle of Goethe's body flies between Alpha Centauri and other stars." Gerhard recalled: "I was intoxicated by this sentence."

Gerhard's father, Julius Schmidt, was the son of a cattle dealer in the small Bavarian town of Baiersdorf. Julius, born in 1872, studied chemistry at the University of Jena under Ludwig Knorr, the discoverer of antipyrene. He graduated in 1894 and six years later began to teach and carry on research at the Technische Hochschule of Stuttgart, the technical university of the State of Württemburg, where he eventually became Professor Extraordinarius.

Julius married Isabella Gombrich, a concert pianist who grew up in Nürmberg, and they established their home in Stuttgart. Gerhard was born December 26, 1901. Three daughters, Elizabeth, Marion, and Renate, followed him, the youngest born in 1924. Gerhard and his sisters grew up in an atmosphere of intellect and arts, in a family that was highly assimilated into German culture but never shook off its Jewish identity. Theirs was a middle-class family, not wealthy—an academic salary was not large—but, like many in the middle class, a family that could live comfortably if modestly, plan for a stable economic life, and look forward to a reliable pension. Well-developed medical insurance contributed to a sense of security. Such a family could embrace a balanced and rewarding life with interests in humanities, arts, and science, with a strong emphasis on education. People showed great respect for someone in the teaching profession; and the combination of research and teaching at a college gave Gerhard's father a prestigious station in life.

Julius Schmidt was a productive scientist, who became well known for both his original research and his publication of an annual review of organic chemistry and a textbook on organic chemistry, *Synthetisch-Organische Chemie der Neuzeit*, which saw several editions and was translated into English. His annual reviews were favorably reviewed in the *Journal of the American Chemical Society*. The British journal *Nature* noted his additional appointment, in 1923, as reader in chemistry at the Engineering College in Esslingen, near Stuttgart. Before the First World War, Julius published at least 140 articles, a substantial portion of them in a leading German chemistry journal, *Berichte der Deutschen chemischen Gesellschaft*. They included a series of more than 30 publications on the chemistry of phenanthrenes and their derivatives, their interconversions and conversions to fluorenes. Phenanthrene is a polycyclic hydrocarbon consisting of three fused benzene rings, a structure that forms the skeleton of natural opiates like morphine. Julius was 42 years old when the Great War began but he served in the German army, on the eastern front, first as Feldwebel, comparable to a senior noncommisioned officer (master sergeant). As a result of an operation in which Russian soldiers were captured, and in a step that was otherwise unusual for a Jew, he was promoted to the rank of an officer, a lieutenant, and was awarded the Iron Cross Medal. The war, however, caused a gap in his scientific output. His publications resumed in 1922, comprising about 50 more articles, some in the direction of applied chemistry, in fabric dyeing and in pharmaceuticals, but also including new reports on the chemistry of phenanthrenes. These articles and the annual reviews and textbook revisions continued through 1932, the year before his untimely death, which also was noted in the journal *Nature*.

Julius and Isabella, Gerhard's parents, remained Jewish and wanted their children to receive instruction and know about their Jewish religion. They looked down on those Jews who converted to Christianity for practical reasons associated with careers. Still, their actual Jewish observance was limited mainly to synagogue attendance at Rosh Hashanah and Yom Kippur services and fasting on Yom Kippur. On those holy days, it was unthinkable to ride a streetcar to the synagogue; one had to walk. For the rest of the year, there was little observance or formal connection to a Jewish community. They did not follow dietary laws and did not celebrate a Passover seder. Their friends were both Jews and non-Jews, as were Gerhard's playmates. While encouraging assimilation into German life, Julius would not have approved of intermarriage. Isabella was more lenient in that respect. Her father had directed a private school in Nürnberg, and served as Gemeinde—president of the Jewish community. Family life for Isabella before she married was much more outspokenly Jewish than was that of Julius; but, although she certainly kept a Jewish household, she was a performing artist and was very liberal minded.

Gerhard himself also grew up as a liberal, assimilated, culturally German Jew. He attended the public Eberhardt Ludwig Gymnasium in Stuttgart, a school with an emphasis on Greek and Latin rather than science. He read a great deal of German literature. He had six years of instruction in French but just one in English, and that involved reading *Julius Caesar*. He developed a love of music and learned to play the cello. Indeed, a cello would accompany him on his many journeys throughout his life; he would take every opportunity to play it, especially in small groups.

Religion was a subject taught in both elementary schools and the gymnasium, with different teachers for different faiths. In Gerhard's gymnasium class in Stuttgart, there were four Jews among thirty-five students in one of the two public schools; there were no Jews in the other parallel school of thirty-five students. The Jewish community arranged for their children's instruction in religion, for which they were separated from the rest of the class. The Chief Rabbi of Stuttgart was their teacher in Gerhard's last year in the gymnasium. In an earlier year, his course schedule did not match that of the other Jewish students so he could not take the religion classes with them; his parents then arranged for a cantor to give him private lessons at home. Although one of his close school friends became an ardent Zionist, whose family emigrated to Palestine, Gerhard gave little thought to the creation of a Jewish state. It seemed unrealistically utopian.

In his youth Gerhard did not experience direct anti-Semitic attacks. On the other hand, the family did not belong to any social clubs, and religious teachers made students aware of the history of Jewish persecution and of limitations for Jews in current German society. In fact, Gerhard felt that, as religious instruction progressed beyond the Bible, there was too much emphasis on the negative history and "the real beauties of fabulous ideas of Jewish philosophy were not emphasized as they should have been."

He encountered expressions of German chauvinism in school, but was impressed with the way in which teachers, whatever their beliefs, maintained an atmosphere of objectivity and tolerance of free expression, even during wartime. During his penultimate year in the gymnasium, when he was 16 or 17, one of the teachers asked the students to write an essay on any subject of their choosing. Based on the books that had so impressed him, Gerhard composed his essay on *Man and His War Against Nature*, including both some biological ideas and ideas of how society should be organized like a biological organism. His teacher handed it back with the marking "Green —complementary to red," intimating both immaturity and coded disapproval of a socialistic tendency. Still, there was no hindrance to his freedom of expression and no attempt to "correct" his thinking. Graduating with distinction, he was valedictorian of his class at commencement exercises in the spring of 1919.

Gerhard was a few months shy of 13 when World War I broke out on the first of August, 1914. "When my father was in the war since starting with 1914 he had practically no influence, of course; he came home only when he was on leave. And then came, of course, defeat in the war. But when I studied—I told my father I wanted to study some kind of natural science—I wasn't quite sure [whether] chemistry, physics, or biology—and he said, 'As a Jew in Germany, if you want to study'—and this was long before Hitler, he had come home from the war of course, it was 1919—'and you are interested in natural science, sure, I understand that completely; but that is a career of employment, you are always an employee of the state in Germany and you will have a very difficult [time] as a Jew to get anywhere, so you should study in order to be independent. You should study medicine, and then you can still see. Besides, medicine is a very interesting branch of natural sciences.'

So I liked that idea and studied it. In Germany you don't go to college; you register for medicine. This is what I did. I studied in Tübingen."

BEGINNING MEDICAL STUDIES IN TÜBINGEN

Tübingen, about 30 miles south of Stuttgart, was a city of about 15,000 people at the time, and the University of Tübingen was the State University of Württemburg. Many students belonged to fraternities, but Gerhard did not join one. His father did not approve of them and, in any case, as a Jew, he would not be accepted into a fraternity. Without this distraction, he devoted all his time to study in the first years of medical school between 1919 and 1922.

Although Tübingen was a most charming town, life in these postwar years was economically difficult, and food was in short supply. Fraternities and their members received support from industrialists, whose children were among the members. Other students depended almost entirely on a food aid program established by Herbert Hoover; Gerhard recalls potatoes, Argentinean corned beef, and cocoa among the "Hoovers."

In his first year of medical studies, he was able, through his father's connections, to rent a room in town; but for the following summer semester he could find accommodation only in Pfrondorf, a village about four miles from Tübingen. There a restaurant rented him a dance hall that was not used during the summer —a large room with twenty-two chairs and no cupboard. He had to walk the four miles to and from the Pfrondorf every day, getting to his morning chemistry class at 8 a.m.

Just as food was in short supply, so was the coal that would be needed to heat his room during the winter. In his first winter he had saved 100 pounds of coal, but had no place to keep it in Tübingen during the summer; it was precious, so he had to carry it out to Pfrondorf at the beginning of summer and carry it back to Tübingen for the next winter.

Gerhard completed five semesters at the University of Tübingen. For the first three semesters, the plan of the lectures was entirely fixed for medical students because they were all compulsory lectures at any medical school: anatomy, with dissection; physiology; inorganic and organic chemistry; physics; zoology; and botany. Biochemistry started in the fourth semester. Up to this point there was no course in pathology and no exposure to patients or clinical teaching of any kind. He attended some lectures outside the medical sciences, in history, calculus, and the national economy, the latter from a very well-known socialist professor at the university. During his first two summer vacations he studied; in the third summer he did volunteer work in the biochemistry laboratory of the agricultural school of Württemberg, near Stuttgart, where Hans Thierfelder had become professor.

Hans Thierfelder was one of several outstanding faculty members at Tübingen. He had studied with the great Felix Hoppe-Seyler in Strassburg and had become professor of physiological chemistry at Tübingen in 1909. His research included important contributions to studies of glucuronic acid and phospholipids, including phosphatides and cerebrosides, subjects that would engage Gerhard much later in his research career; and, with Hoppe-Seyler, he wrote a handbook of chemical analysis. One of Thierfelder's students, who received his doctorate in 1923 and so was present during Gerhard's studies, was Ernst Klenk,[5] who continued at Tübingen until 1936 and then went on to become professor of physiological chemistry at the University of Köln. He discovered cerebrosides in brain and phosphatide deposition in Gaucher's and Niemann-Pick's diseases. Another notable professor in physiological chemistry, Franz Knoop, discovered beta-oxidation of fatty acids, reported in a classic paper in 1904.[6] Gerhard recalls his favorite teacher was Friedrich Paschen, a spectroscopist who was professor of physics.[7]

Gerhard became interested in biochemistry. He was the only medical student to take up an elective laboratory experience. A very kind associate professor, Fritz Wrede, accepted him, a student without any laboratory experience, into the lab and guided him in biochemistry exercises. Wrede needed some amino acids, so he had Gerhard look up papers by Emil Fischer and follow Fischer's procedure for amino acid preparation by the Dakin extraction, with butyl alcohol.[8] Some of the preparations did not go smoothly, but Gerhard enjoyed these exercises enormously.

An experiment that most distinctly did not go smoothly occurred when he worked in the Thierfelder lab. On one occasion, Thierfelder had him prepare and crystallize the protein edestin from hemp seeds. "He gave me practically a pound of hemp seed to extract with petroleum ether.[9] And I put it into a four-gallon glass jar with a glass stopper—that big. Of course, I mean, I didn't have any experience with this—this petroleum extraction. I put a kilogram of hemp seeds in this jar, stoppered it, and after two minutes this glass [exploded and] broke one lamp in the laboratory. It was a great tragedy; our only glass-stoppered jar."

[5]G. Klenk and H. Faillard, "Professor Ernst Klenk: Biography and Scientific Achievements," *Journal of American Oil Chemists' Society* 43(1966): 48A.

[6]F. Knoop, "Der Abbau aromatischer Fettsäuren im Tierkörper," *Beiträge zu chemische Physiologie und Pathologie* 6(1904): 150.

[7]In 1889 Paschen had described an equation for the unexpected way in which the voltage required to create an electric arc between metal plates varies with the distance between the plates and the atmospheric pressure; "Paschen's Law," in "Über die zum Funkenübergang in Luft, Wasserstoff und Kohlensäure bei verschiedenen Drucken erforderliche Potentialdifferenz," *Annalen der Physik* 273(1889): 69–75.

[8]H. D. Dakin, "On Amino Acids," *Biochemical Journal* 12(1918): 290.

[9]The extraction with petroleum ether is a preliminary step to remove lipids before application of the Dakin procedure, as in G. Reeves, "XLVIII A new method for the preparation of the plant globulins," *Biochemical Journal* 9(1915): 508.

Gerhard was not completely satisfied with the teaching of biochemistry at Tübingen. The courses dealt with chemical composition of tissues and body fluids and aspects of nutrition, but not with biochemical mechanisms or new developments in intermediary metabolism, the chemical changes involved in processes like oxidation of foodstuffs or the synthesis or breakdown of chemical constituents of the tissues. There was no indication to the students that one could learn something about human processes by studying related processes in bacteria or yeast.

Once again, a book encounter became an important determinant of his career path. He frequented a very good bookstore at the university and had bought and studied the main biochemistry textbook, written by Olof Hammarsten, an excellent book for descriptions of chemical properties and preparations.[10] But then, in his last semester, there appeared a book written by Ernst Schmitz, a professor in Breslau,[11] who had been a student of Gustav Embden in Frankfurt. Schmitz's book on biochemistry put intermediary metabolism in the foreground. At that time, most of the knowledge on intermediary metabolism came from liver perfusion experiments, and Embden's lab was a prominent contributor, with studies on amino acid transformations in liver as well as glucose breakdown to compounds like lactate and pyruvate (glycolysis).[12] Using liver perfusion, Embden had also shown, by 1910, the biological formation of acetoacetate from acetate.[13] This kind of biochemistry was not taught at Tübingen. The book by Schmitz opened a new world for Gerhard, leading him to decide to transfer to Frankfurt for his years of clinical studies if he could somehow become associated with Embden's lab.

He was so stimulated by the Schmitz book that when the professor of physics in Tübingen, Professor Paschen, boasted of making the best galvanometers in the world so that even the great English physiologist A.V. Hill bought his galvanometers from Paschen, Gerhard decided to look into papers published by Hill and he was further excited by the physiological chemistry in those works. Struggling with the English, he nevertheless was able to learn from them. Then "I had the nerve to—I attended, also, a seminar in zoology in the Zoology Department; and I announced a seminar in the Zoology department

---

[10] Olof Hammarsten, *Lehrbuch der physiologischen Chemie* (Weisbaden: J. F. Bergmann Verlag, 1893) and subsequent revisions, which were translated into English by John A. Mandel, for example, *A Text-book of Physiological Chemistry*, 5th American edition (New York: John Wiley & Sons, 1909).

[11] Ernst Schmitz, *Kurzes Lehrbuch der chemischen Physiologie* (Berlin: Karger, 1921).

[12] In the procedure of liver perfusion, the liver and its major blood vessels are isolated and suspended in a physiological saline solution. Substances of interest are introduced into the large blood vessels entering the liver, and fluid is collected from the main blood vessel exiting the organ. The procedure allows one to determine which products are formed as the fluid passes through the liver cells, between entry and exit, as a result of cellular action on the substance that was introduced.

[13] G. Embden and J. Wirth, "Inhibition of Acetoacetic Acid Formation in the Liver,"*Biochemische Zeitschrift*, 27(1910): 1–19.

as a preclinical medical student, entitled "The significance of lactic acid formation as a source of energy for muscle contraction" and it was greatly admired by the professor of zoology, who knew of biochemistry just as much as I would know about the classification of worms."

At Tübingen, Gerhard completed his "physicum," the examination of all the medical sciences—anatomy, chemistry, physics, zoology, botany, biochemistry, and physiology. He wrote a letter to Gustav Embden, letting him know he was excited by what he had read and that he had worked as a volunteer in Thierfelder's lab, and asking whether he might do something like that while he continued with clinical studies at Frankfurt. Embden wrote a welcoming reply.

It was quite usual for medical students to transfer from one school to another, often motivated by interest in a renowned professor at a given university. As curricula were compulsory and largely interchangeable at different schools, it was a simple process to move from one school to another; some students would attend four or five different universities during their medical studies. One motivation was the attraction to certain cities because of their interesting cultures. More important, however, students chose to attend a university, even if for just a while, to learn from particular professors. One would go to Berlin for Emil Fischer or to Tübingen to hear Franz Knoop. As for clinical people, they went to Strassburg, for example, to hear Minkowski who, with Joseph von Mering, had demonstrated in 1889 that diabetes developed in dogs after their pancreas was removed surgically.[14] It was partly because of financial limitations, in dire times, that Gerhard moved only once in his medical school career. Of course, it was largely because of Gustav Embden that he settled into Frankfurt and stayed.

---

[14] Joseph von Mering and Oskar Minkowski, "Diabetes mellitus nach Pankreasextirpation," *Centralblatt für klinische Medicin, Leipzig,* 10(1889): 393; *Archiv für experimentelle Patholgie und Pharmakologie, Leipzig,* 26(1890): 37.

Two

Weimar Years:
These Are the Things I Remember

Gerhard moved from Tübingen to Frankfurt in 1922, turbulent times that followed the First World War, with the fledgling Weimer Republic under siege from both the left and the right, and with economic instability leading to the devastating hyperinflation of 1923. Though he focused largely on his studies, his life was profoundly affected by events surrounding him. He recalled several major issues.

In these postwar years Gerhard became aware of anti-Semitism in ways he had not experienced before. "The early signs were that, with the general decay of, we could say, Victorian formalities in life after the war, the expressions of political opinion became more and more vulgar. . . . With the unfavorable economic developments after the First World War, many political parties looked for a scapegoat to blame; and the Jews—not only in Germany—were a fabulous scapegoat for such expressions."

Political parties with frankly anti-Semitic platforms had campaigned in Germany since late in the nineteenth century, and conservative parties expressed anti-Semitism even when it was not part of their official platform;[1] but growing anti-Jewish sentiment expanded late in the Great War, when chauvinistic pro-war nationalists charged that Jews contributed to armistice-seeking initiatives and that Germany lost the war because she had been stabbed in the back by internal forces, Jews prominent among them. In response to the armistice initiative, a pro-war German nationalistic Fatherland Party was formed in 1917.[2] That party was dissolved after the 1918 revolutions, but in the next year one of its members, Anton Drexler, led formation of the German Workers Party, which later became the National Socialist Workers Party. Until the end of the war, Gerhard recalled, expressions of anti-Semitism in most walks of life were heard but were muted by a general code of civility.

---

[1] Amos Elon, *The Pity of It All, A History of Jews in Germany, 1743–1933* (New York: Henry Holt, 2002); Ismar Schorsch, *Jewish Reactions to German Anti-Semitism, 1870–1914* (New York: Columbia University Press, 1972).

[2] The events and people in this chapter are those recalled by Gerhard during interviews. Details are filled in based largely on R. J. J. Evans, *The Coming of the Third Reich* (New York: Penguin Press, 2004).

With defeat in the War, the expression of anti-Semitism, as of all political differences, grew more vulgar, and discourse often was displaced by violence.

Moreover, after the war, anti-Jewish and anticommunist feelings reinforced each other, and anti-Semitic ultranationalists made Jews and communists their common targets. Although the vast majority of Jews were politically moderate, voted with Social Democrats and centrist parties, and loathed communism, some Jews had leading roles in the short-lived radical socialist– communist Spartacist uprising in Berlin at the beginning of January 1919. The uprising was quashed fiercely by January 12 (with assent of the new and still unstable Ebert government) by the paramilitary Freikorps (Free Corps), a group of private armies formed by former senior army officers. The Spartacist leaders, Rosa Luxemburg and Karl Liebknecht, both Jews, were assassinated on January 15, 1919. At about the same time, Kurt Eisner, a Jew, had declared and led a short-lived Bavarian Socialist Republic on November 7, 1918; he was assassinated on February 21, 1919, by a fanatical nationalist aristocratic student, Anton Graf von Arco. In the confusion following the assassination, a Social Democrat cabinet tried unsuccessfully to rule in Bavaria, and a communist leader of Russian Jewish origin, Eugen Levine, proclaimed a Bolshevik regime in Munich.

By March, a Red Army assembled by the communists was routed in an invasion by 30,000 troops of the Free Corps, who established a "White" counterrevolutionary government in Munich. Wolfgang Kapp, a founder of the Fatherland Party, led a 5,000-member force of the Free Corps in an attempted government takeover in Berlin, in March 1920, forcing the elected leader of the Weimar Republic, Friedrich Ebert, to flee temporarily. Heeding a call from the government, however, workers across the country did not accede to the Kapp forces' demands and went on a general strike. The putsch attempt was defeated. Kapp fled to Sweden.

Jews who had prominent government positions in the Social Democratic administration of the Republic were also targets of right-wing assassins. Walter Rathenau, a Jewish industrialist, politician, and writer, whose father was also a major industrialist, served as Foreign Minister in the Weimar government. He was assassinated June 24, 1922, by a group of Free Corps members. In all, more than 200 political assassinations were committed by rightist groups by 1923.

Ominously, in the response to violence and assassination, the very conservative judiciary was an influence that reinforced the radical nationalist forces, exonerating or giving only brief imprisonment to a large majority of right-wing offenders accused of murder, while meting out more severe judgments and sentences, including execution, against a large fraction of leftists. The assassins of Luxemburg and Liebknecht, for example, received light sentences.

Eisner's assassin, Arco-Valley, was originally sentenced to death in 1920, but a conservative judge eventually reduced this decree to five years in prison. While two of Rathenau's three assassins committed suicide, one was captured; he received a sentence of 15 years, but was released after less than 5 years.

By 1921, Adolf Hitler, with the support of General Ludendorff, had entered and risen to take over the German Workers Party (re-named the Nationalist Socialist German Workers' Party in 1920), a successor to the Fatherland Party. He organized arch-nationalist anti-Semitic forces, including paramilitary storm troopers. By the fall of 1923, inspired by Mussolini's march on Rome a year earlier, he thought he had enough power to launch a coup from Munich—the Beer Hall Putsch—aiming to go onward to Berlin to replace the Republic's government. His putsch attempt was initiated during the evening of November 8, in a political assembly at which the Bavarian Commissioner, Gustav von Kahr, was speaking. Storm troopers surrounded the hall, while Hitler, with some twenty colleagues, among whom were Herman Göring and Rudolph Hess, entered, announced the coup, sequestered von Kahr, and told the assembly that von Kahr was supporting him in his plan to march on to Berlin; however, a police force blocked the marching storm troopers the next day and the putsch was crushed. Its perpetrators, though having committed acts of treason, were treated lightly by the judiciary. Ludendorff was acquitted. Other leaders were found guilty but released. Hitler served only nine months of a five-year sentence. Held at Landsberg am Lech, a comfortable prison, he received visitors and composed his manifesto *Mein Kampf* during his detention. These were the examples of unrest that stood out in Gerhard's memory.

In addition to these events and persons of political history, Gerhard recalled the story of the Nobel Prize-winning Jewish chemist, Richard Willstätter,[3] as one of the strong signals of anti-Semitism in academia. Willstätter had studied with the renowned chemist Adolf von Baeyer,[4] in Munich. He earned his doctorate in 1894, determining the structure of cocaine, and continued at Munich until 1905. "It was almost a rule that the first man in Baeyer's department became professor in Zurich in Switzerland. And then, from there, if he was really outstanding he was called back to Germany." So it was with Willstätter, one of Baeyer's outstanding students. While in Zurich, he determined the structure of the plant pigment chlorophyll, for which he was awarded the 1915 Nobel Prize in chemistry. Meanwhile, when the Kaiser

---

[3] *Nobel Lectures, Chemistry 1901–1921* (Amsterdam: Elsevier, 1966). Retrieved 21 Sept 2012: "Richard Willstätter—Biography." Nobelprize.org. http://www.nobelprize.org/nobel_prizes/chemistry/laureates/1915/willstatter-bio.html.

[4] *Nobel Lectures, Chemistry 1901–1921* (Amsterdam: Elsevier, 1966). Retrieved 21 Sept 2012 "Adolf von Baeyer—Biography," Nobelprize.org. http://www.nobelprize.org/nobel_prizes/chemistry/laureates/1905/baeyer.html.

Wilhelm Institutes (now known as the Max Planck Institutes) were founded in 1912, Willstätter was invited to return to Germany as director of the Kaiser Wilhelm Institute for Chemistry in Berlin–Dahlem. Jews were much more likely to be accepted as leaders at these Kaiser Wilhelm Institutes than as members of university faculties, but because of his wide recognition, Willstätter was also appointed professor of chemistry at the University of Berlin.

When von Baeyer stepped down, the chair of chemistry at the University of Munich became open. "Of course, the decision to invite somebody to take over the chair was made by the Secretary of Education of the state in question in Germany. The faculty in Munich has to propose three candidates for the chair, and the minister, the Secretary of Education, decides which one should receive the call. And the Secretary of Education decided Willstätter. And when Willstätter had accepted the invitation, the king of Bavaria has to sign the final contract. When the Secretary of Education presented this document to the king for signature, the king told him, 'Das ist aber das letzte Mal, daß ich Ihnen einen Juden unterschreibe.' 'That's the last time that I again sign up for a Jew.'" At Munich, Willstätter undertook pioneering research on enzymes and their mechanisms, although he did not accept the idea that catalytic activity was an intrinsic property of proteins. After several years, when a faculty position for a crystallographer became open in his department, Willstätter strongly recommended Victor Goldschmidt,[5] a professor in the Mineralogy Institute of the Christiana University (Oslo University), to fill it. Goldschmidt, born of Jewish parents in Zurich, had obtained his PhD in Oslo and had determined the crystal structure of many inorganic compounds; he is considered a founder of the field of geochemistry. As Gerhard recalls, based on reading Willstätter's autobiography, "A number of faculty members disagreed with Willstätter and quite openly mentioned they wouldn't like to have another Jew on the faculty." In response to increasing anti-Semitism, Willstätter resigned from his chair in 1924 and never took up another one, though he received offers both in Germany and abroad. He fled the Gestapo in 1938 and moved to Zurich, where he died in 1942.

With direct knowledge of the limitations for Jews in academia and in government service, Gerhard's father, Julius, added a new element to his previous advice that Gerhard should study something like medicine, a profession in which one had an opportunity to be independent of state-determined support. By 1926, at the end of Gerhard's medical studies, his father warned him further: "Now you are making plans for your career, it would be very

---

[5] C. C. Gillespie and N. Koertge, eds., "Goldschmidt, Victor." *Complete Dictionary of Scientific Biography* (New York: Charles Scribner's Sons, 2007 [ebook]). Retrieved September 21, 2012, from Encyclopedia.com: http://www.encyclopedia.com/doc/1G2-2830901678.html.

wise not to plan too much for staying in Germany although I know that you would like to do that." Julius showed a realistic reading of ominous signs.

Gerhard's interviewer, Morris Cynkin, sought to understand why Jews continued to live in Germany in spite of seeing this series of threatening events in the early 1920s. Gerhard responded that there were still many very enjoyable features of living—and studying—in Germany. Jews were highly assimilated into German culture and the Weimar years produced an incredible burst of creative expression in science, literature, art, music, theater, architecture, and industry. Notably, there was a free and diverse press, with outstanding liberal newspapers and magazines published in the major cities, where the majorities of Jews lived. Although one was always conscious of the presence of anti-Semitism, "You just put up with it because the positive sides, I believe, the attractive sides of living in Germany at that time, were so appealing to, certainly, most of the Jews that they put up with the other side as a necessary disadvantage." As pointed out by Michael Brenner, the Weimar period also saw a burst of Jewish cultural expression, a virtual renaissance and return to Jewish identity by many Jews.[6] On the balance, as the rumble of German nationalism and anti-Semitism was not yet a direct threat on a wide scale, and Jews could not conceive of what its eventual culmination would become, Germany was still a good place to be; the price of living with restrictions and some bad language and some instances of violence by fringe elements of society was acceptable.

---

[6] Michael Brenner, *The Renaissance of Jewish Culture in Weimar Germany* (New Haven, CT: Yale University Press, 1996).

Three

Completing Medical Studies in Frankfurt

The major motivation for Gerhard to move to Frankfurt was the possibility of having some time, even while engaged in his clinical medicine studies, to work in Gustav Embden's lab. For this to happen he did need Embden's approval even though no formal application was required for changing from Tübingen to Frankfurt for medical studies. If Embden had not written that he would accept him, Gerhard would not have made the move. Embden later told Gerhard that, in fact, he had accepted him with some reluctance, perhaps based on what he learned when he wrote to faculty members who served as references relating to Gerhard's years in Tübingen. There, for example, Gerhard had volunteered for some spare-time work in the lab of the physiologist Professor Wilhelm Trendelenburg, who gave him a problem of determining alveolar carbon dioxide in frog lungs. For this "I had to puncture the pleura with a mercury-sealed gas-tight pipette and then very slowly draw a sample of air and then inject this air into a micro metabolic gas analysis apparatus (micro-Haldane) and determine the carbon dioxide. Of course, I spilled some mercury; as you can imagine, it was my second or third semester and I had no idea of instrumentation yet." So when Embden wrote to Trendelenburg to inquire about Gerhard, he may have been told that Gerhard was given a simple experiment and "as a consequence the laboratory was flooded with mercury!" Still, Embden accepted Gerhard's proposal and "he was incredibly nice to me."

Frankfurt was attractive from another point of view as well. The university was founded in 1914 by wealthy Frankfurt citizens, including Jewish business-men and industrialists, and received significant support from the City of Frankfurt, whereas other universities were institutions of the German states. The University of Frankfurt seemed to be more liberal, a more favorable environment for a Jewish student. Embden's department was part of the Theodor Stern Haus, which included physiology and pharmacology depart-ments of the university, as well as a research institute for colloid chemistry and another for x-ray science. The money for this large medical research institution was given by the widow of an industrialist for whom it was named (as is the long riverside road, Theodor-Stern-Kai, on which the university's medical faculty is still located).[1]

Gerhard pursued the clinical portion of his medical studies between 1922 and 1924. During this time he did not have a lot of time in the lab; and, because of his schedule of clinics and lectures, he was not able to audit Embden's lectures. An assistant was assigned to supervise him during the one afternoon of the week available to him in the lab. There he met with one great disappointment. Based on the excitement he felt in reading the book by Ernst Schmitz, one of the methods he was most interested in learning was liver perfusion, through which Emden had contributed so much to knowledge of intermediary metabolism. On asking one of the assistants, Gerhard learned that they hadn't done any liver perfusion in the Embden lab in recent years. Embden's current interest in 1922 was in measurements of inorganic phosphate formation in muscle. "All they did in Embden's lab was snip muscles, snip frog muscles" and measure phosphate.

The cost of living had increased since the end of the war, but was affordable in mid-1922, when Gerhard came to Frankfurt. Soon, however, the great inflation arrived. Living costs increased more and more steeply. "It became so expensive that the amount which I could get from home wasn't enough anymore, so Embden arranged for me that I could stay at the country house of his brother-in-law, a high official in a big chemical company in Frankfurt, the Gold and Silver Company. I had to do a little housework, chop wood, feed the dogs, and take care of the dogs, and shine the shoes of the family, and things like that, and for that the landlord paid for the railway. I had to take a commuter train to Frankfurt every morning. I paid the railroad fare and then I could have dinner in my room without paying." Gerhard recalled that his father had always advised him to save whatever was left from his monthly allowances, and to go to the library rather than buy a book. During the inflation period, however, Gerhard, was glad that he had bought that book or filled other wants, because if he had saved the money, it would have just become worthless.

In January 1923, in response to German defaults on payments of the Versailles Treaty reparations, French and Belgian troops occupied the industrial Ruhr Valley, a major source of coal for Germany. Germans responded with passive resistance and strikes. Railroads ceased operating. "The French closed the border from one night to the next, on twenty-four hours' notice, so I had to move to Frankfurt, which I did. This home where I lived, this village, was in something like the White Mountains here but not as high; near Frankfurt is a beautiful hilly wooded zone, and I had to leave this comfortable home

---

[1] U. Flaig, "Gustav Embden (1874–1933) und die Frankfurter physiologische Chemie." PhD diss., Frankfurt am Main, 1992.

with all my belongings in a knapsack, because the trains—I couldn't use them anymore—they wouldn't let them through—my cello in my left hand. It was a ten-mile hike, fifteen kilometers."[2]

Fortunately, for the rest of that semester and during the vacation, Gerhard was able to stay in Frankfurt, at the home of another brother-in-law of Embden, an assistant professor of pathology. That family was pleased to have someone look after their apartment while they were on vacation in the Bavarian Alps. At the end of vacation, still another accommodation for Gerhard was arranged by Embden, in the home of a close friend, a painter whose wife was the daughter of a very famous professor of comparative zoology in Freiburg. They owned a house at the central square of the old city—the Römerberg. "I lived directly under the roof and had a beautiful view of Frankfurt."

Inflation became hyperinflation. A U.S. dollar cost 1,000 marks in August 1922, 17,000 marks in January 1923, 353,000 marks by April, 4,621,000 marks in August, and 4,200,000,000,000 marks by December.[3] Incredible amounts of money were printed to pay salaries, but the German currency collapsed totally during 1923 and living conditions were exceedingly difficult for many, especially for anyone on a fixed income like a pension. For Gerhard it meant that, at one point, the amount of money he received from home for a month's expenses soon could buy only a can of sardines. Economic life began to turn around only after a new agreement on reparations payments, in the fall of 1923, and new currency policy the following year.

Still, life and study were possible, and Gerhard completed the required clinics and lectures in Frankfurt in 1924. He needed to fulfill two more steps to earn the medical doctor degree: an internship and preparation of a thesis. For the internship he went to the municipal hospital in Stuttgart, so he was able to live at his parents' home; and, as an intern, he had free meals at the hospital. He was accepted at this hospital, partly because of his laboratory experience, by an outstanding professor, Professor Cahn, a patriarchal figure who had moved from Strassburg[4] to Stuttgart to avoid becoming French when Alsace was ceded to France after the war. Cahn had a modern outlook and introduced the new micro method for glucose determinations in the clinical labs. Gerhard was familiar with the method from its use in Embden's lab, and performed a service by doing these measurements, particularly for patients in the Cancer Hospital.

---

[2] The border of French occupation in the Rhineland was very close to Frankfurt.

[3] R. J. Evans, *The Coming of the Third Reich* (New York: Penguin Press, 2004), 105.

[4] As part of Germany before World War I, the city was known as Strassburg; when ceded to France after the war, its name became Strasbourg. The names will be used in accord with the timing of events.

Four

Gustav Embden, a Model Mentor

Gerhard did not have to spend a full year of internship in Stuttgart. An intern was allowed three months of elective training, which could even be laboratory training, and for that reason he returned, in 1925, to Embden and Frankfurt, where he also continued with research needed for his medical-degree thesis. Embden arranged for him a small fellowship from the Rockefeller Foundation. Furthermore, on Gerhard's first visit to his office on returning from Stuttgart, Embden said to him: "You have a fellowship but it's not enough that you can even rent a decent room. With that, would you mind getting a room in the department?" Gerhard quickly accepted the offer. Embden immediately picked up the phone, called the administrator of the hospital and asked him to prepare a room and have in place, by that evening, a bed and night table, bookshelf and linens. Gerhard was indelibly impressed with the fact that, without having to fill out forms or get permission, Professor Embden was able, in a few minutes, to save Gerhard the expense of renting a room. On the downside, the room was in the basement and it had an ice chest, the only cooling facility in the department, and it was right next to the frog room. "There were about three hundred of them. You can imagine in the spring . . . a tremendous concert every evening."

Embden's sensitivity to Gerhard's economic status continued through later stages of the young scientist's training. Even in difficult times, Gerhard managed to accumulate some savings so he could undertake a hiking trip into the Swiss Alps during his summer vacation. One summer, as Gerhard was taking his leave, "Embden asked: 'Going again to Switzerland? . . . You work quite hard. I think you deserve having a good time in Switzerland.' He gave me 200 marks, from some funds of the department, to spend for my vacation." Gerhard recalled that Embden had some discretionary funds even though fellowships were small and life was generally austere in Germany, while the country was paying World War I reparations stipulated by the Treaty of Versailles. Embden had a large laboratory, built by the University of Frankfurt, which received generous contributions from private sources.

From the beginning of their relationship, Gerhard considered Embden to be a very special and caring mentor. Years later, he said Embden stood out above every other person with whom he worked in his manner of teaching and consideration of his students' needs.

Gustav Embden, who was 51 years old when Gerhard moved back to Frankfurt to work on his medical-degree thesis in 1925, was a grandnephew of Heinrich Heine and son of a well-known lawyer in Hamburg.[1] During and after medical studies, between 1897 and 1904, Embden had received research training in Strassburg with Franz Hofmeister,[2] head of the Physiological Chemistry Institute and an early pioneer in metabolic chemistry. Hofmeister trained many outstanding scientists, including Franz Knoop (β-oxidation of fatty acids), Otto Loewi (urea synthesis and acetylcholine as neurotransmitter), and Alexander Ellinger (pharmacodynamics), as well as Gustav Embden.

In his years in Strassburg, Embden undertook research with liver perfusion, in which he introduced known substances into blood vessels entering the isolated liver and measured their transformations as reflected by analyses of substances in blood vessels leaving the liver. He thereby measured metabolic chemical changes in carbohydrates, amino acids, and fatty acids. The significance of Embden's early research results with this method was noted by Carl von Noorden, hospital director and head of the medical clinic at City Hospital Sachsenhausen in Frankfurt. At von Noorden's invitation, Embden became an assistant in the hospital and its laboratory in Frankfurt in April 1904.

Embden became part of the faculty of the University of Frankfurt in the year of its foundation, 1914. There he established the Division of Vegetative Physiology, to which Gerhard Schmidt came for his research training. Gerhard found it interesting that Embden always insisted that he was studying the physiology of cells rather than biochemistry. The term "vegetative physiology" distinguished this cellular metabolic division from that of animal physiology, which studied more traditional aspects of physiology, such as function of the neural, muscular, cardiovascular, and other systems.

Embden's work with liver perfusion yielded major contributions to knowledge of the metabolic changes, including the demonstration, published in 1906–08, that certain amino acids and fatty acids with even numbers of carbon atoms were converted to acetone and acetoacetate. With Ernst Schmitz, he demonstrated the formation of amino acids in liver: alanine from lactate and tyrosine from its keto acid (1911).[3] This Ernst Schmitz was the author of the book that motivated Gerhard to move from Tübingen to Frankfurt. It

[1] For biographical information on Embden, see Ulrich Flaig, *Gustav Embden (1874–1933) und die Frankfurter physiologische Chemie* (Frankfurt am Main: Senkenbergischen Institut für Geschichte der Medizin der Johann Wolfgang Goethe-Universität, 1992), 3–13; L. Jaenicke, "Eine kleine Geschichte des Emben-Meyerhof-Zyklus: Gustav Embden und die vegetative Physiologie," *Biospektrum, Springer Spektrum* 2(2000): S129; D. Nachmansohn, *German-Jewish Pioneers in Science, 1900–1933* (New York: Springer-Verlag, 1979), 327–36.
[2] J. L. Abernethy, "Franz Hofmeister: The Impact of His Life and Research on Chemistry," *Journal of Chemical Education* 44 (1967): 177.
[3] G. Embden and E. Schmitz, "Über synthetische Bildung von Aminosäuren in der Leber," *Biochemische Zeitschrift* 29(1910): 423–8.

was, in particular, Schmitz's descriptions of these liver perfusion experiments that provided Gerhard's strongest stimulus.

Focusing on carbohydrate degradation, Embden discovered pieces of the puzzle of how glucose or glycogen is converted to lactate in animals or ethanol in yeast, a process that came to be known as "glycolysis." He would eventually solve the core of the puzzle in 1932 with creative insight, based on extensive research in his and other laboratories, notably those of Arthur Harden in England; Otto Meyerhof in Kiel, Berlin, and then Heidelberg; and Jacob Parnas, a Polish biochemist who had also studied with Hofmeister. The process of glycolysis became known as the Embden–Meyerhof or, less widely, as the Embden–Meyerhof–Parnas pathway.[4]

During World War I, an important change in direction was approaching for Embden; he was turning away from study of the liver, toward investigation of lactate formation in muscle tissue. With muscle, an important issue was how the mechanical energy of muscle contraction was produced from the chemical energy, and which substances provided that energy. The great English physiologist A. V. Hill had measured the energy involved in muscle contraction, and in 1912 Otto Meyerhof, then in Kiel, found that energy was nearly identical with the energy released by chemical conversion of glucose to lactate during the contraction. For these insights, Hill[5] and Meyerhof[6] were awarded the 1922 Nobel Prize in Medicine and Physiology. Thus a lot of attention was focused on formation of lactate as a possible direct source of the energy. Embden was not convinced of such a direct linkage. This was one of the differences between Embden's and Meyerhof's interpretations of the rapidly increasing information about the chemical physiology of muscle.[7]

Embden aimed to identify the natural precursor of lactate in muscle, without addition of an exogenous substrate, such as glucose. In 1914, he proposed that there was an endogenous precursor of lactate, a compound he named "lactacidogen" because he could partially purify a substance, from a tissue extract, that could give rise to lactate and inorganic phosphate when it was added to muscle juice.[8] Over the next 14 years, he published several papers

[4]O. Meyerhof, "Intermediate Products in the Last Stages of Carbohydrate Breakdown in the Metabolism of Muscle and in Alcoholic Fermentation," *Nature* (1933): 132: 337–40, 373–75; C. F. Cori and G. T. Cori, "Carbohydrate Metabolism," *Annual Review of Biochemistry* 3(1934): 151–74; J. A. Barnet, "A History of Research in Yeasts 5. The Fermentation Pathway," *Yeast* 20(2003): 509–43.

[5]*Nobel Lectures, Physiology or Medicine 1922–1941* (Amsterdam: Elsevier, 1965), "Archibald V. Hill— Biography," retrieved 29 Sept 2012, from www.nobelprize.org/nobel_prizes/medicine/laureates/1922/hill.html

[6]*Nobel Lectures, Physiology or Medicine 1922–1941* (Amsterdam: Elsevier, 1965), "Otto Meyerhof— Biography," retrieved 29 Sept 2012, from www.nobelprize.org/nobel_prizes/medicine/laureates/1922/meyerhof.html

[7]David Nachmansohn, *German-Jewish Pioneers in Science, 1900–1933: Highlights in Atomic Physics, Chemistry, and Biochemistry* (Berlin: Springer-Verlag, 1979), 333.

[8]G. Embden, F. Kalberlah, and H. Engel "Über Milchsäurebildung im Muskelpressaft.1. Mitteilung," *Biochemische Zeitschrift* 45 (1912): 45–62.

on the chemical behavior of lactacidogen and its formation and degradation. He did not commit to its precise identity, but considered it to be a substance containing one or more phosphorylated sugars. In the mid-1920s, however, it was still not clear whether the several known phosphorylated sugars were direct participants in a physiological chemical pathway from glucose to lactate.

By the time Gerhard reached the laboratory in 1923, Embden had become less interested in metabolic chemical conversions and was exploring the hypothesis that muscle contraction involved reversible changes in the "colloidal state," that is, the physical size, solubility, and shape of muscle proteins, and that ions were effectors of these changes.[9] This proposal probably reflected his training with Hofmeister, who had described the classification of ions, with an increasing order of effectiveness (the Hofmeister series of ions), in terms of their effects on protein "colloidal properties." Proteins became "swollen" with uptake of water in low ionic concentration and then, with higher ion concentrations, they decreased in volume and became insoluble.

Embden had shown earlier that addition of fructose-1,6-bisphosphate as a substrate to muscle juice gave rise to both lactate and inorganic phosphate and that, in the short term, there was a one-to-one quantitative relationship in the formation of a molecule of lactate for each molecule of inorganic phosphate.[10] He considered the possibility that an increase in inorganic phosphate concentration was affecting the shape and colloidal properties—the contraction—of muscle protein.

Gerhard recalled: "And so from the time that I came to help at this laboratory, he was tremendously interested in the effects of ions on the lactic acid and phosphate formation. The whole laboratory did nothing else but phosphorous determinations. At that time every collaborator got as his dowry, so to speak, for his laboratory work, an enormous dessicator and a battery of suction flasks and a manifold to put under the pump, because Embden had, just in 1921, developed an excellent gravimetric micro-method for phosphate determination using strychnine molybdate.[11] And all the phosphate determinations in the laboratory were done by this method—gravimetric. . . . The whole department just systematically tried out every metal ion and every anion under the sun with minced frog muscle."

---

[9] G. Embden and H. Jost, "Über chemische und kolloidchemische Veränderung bei der Muskelermüdung und ihren biologische Zusammenhang,"*Zeitschrift für physiologische Chemie* 165 (1927): 224–54.

[10] Only later was it learned that the inorganic phosphate did not come directly from the steps between fructose-1,6-bisphosphate and lactate, but rather from the hydrolysis of ATP (adenosine triphosphate) that was formed during those steps. See David Nachmansohn, *German-Jewish Pioneers in Science, 1900–1933: Highlights in Atomic Physics, Chemistry, and Biochemistry* (Berlin: Springer-Verlag, 1979), 329–31.

[11] Embden's method, described in G. Embden. "Eine gravimetrische Bestimmungsmethod für kleine Phosphorsäuremengen," 113 (1921): 138–45, was based on a gravimetric method first described by I. Pouget and D. Chouchak, "Dosage colorimétrique de l'acide phosphorique," *Bulletin de la Société Chimique de Paris* series V(1909): 104 and series IX(1911).

Not surprisingly, Gerhard's medical thesis was on "Colloid chemical changes in muscle of rabbits poisoned with strychnine," and his first publication, in the journal *Arbeitsphysiologie*, in 1928, was a 17-page article entitled "Über Kolloidchemische Veranderung bei der Ermüdung des Warmbluter-muskels" (On colloidal chemical changes in fatigue in muscle of warm-blooded animals). In this article, Gerhard reported that there was a more rapid increase of inorganic phosphate, reflecting glycogen breakdown, in strychnine-treated rabbits than in untreated animals. Embden was interested in the effect of strychnine because he had shown earlier that the lactacidogen–phosphate complex is decreased in muscles as a result of rapid repeated contractions caused by strychnine.

Gerhard completed his thesis research and earned the medical degree, with distinction, in 1926. "In Germany, you cannot call yourself M.D. unless you submit a thesis. You can either make a work on a collection of 100 cases of this or that disease and have it printed on your own expense and submit that to the faculty and then you get your degree, or you can do experimental work and if something publishable results, you are excused from having your results printed. You can just order so many reprints and give that to the faculty as your thesis. This is one reason why many medical students went to a theoretical laboratory. It was not the reason for me. My thesis was in Embden's laboratory."

At this time the main junior faculty members in Embden's department were Emil Lenhartz, Hans J. Deuticke, and Hans Jost, all three of whom played significant roles leading to Embden's great insight into the chemical process of glycolysis, a critical initial phase in the chemical breakdown of the sugar glucose in living tissues and the capture of its energy and material for sustaining life and building diverse substances needed by living cells. Jost, sharing characteristics for which people from the Rhineland were known, was more outgoing than the others. Gerhard recalled that Jost was very gracious to him, introducing him to new social experiences, including his first exposure to a Frankfurt wine restaurant and to an alcoholic beverage—and therein his first experience of becoming quite drunk. As they became friendly, Gerhard felt free to express his liberal political views, but was warned by other faculty members to be cautious in his speech because Jost was considered likely to become a member of the Nazi Party. Eventually Jost, Deuticke, and Lehnartz all did so, and all retained academic posts during the 1933–45 period.

A NEW TOPIC FOR PRODUCTIVE HABILITATION RESEACH

Gerhard was not particularly interested in the subject of his thesis research; it just happened to be what Embden asked him to work on. Embden was quite satisfied with his work, which, at Gerhard's own initiative, provided more detail on the time course of reactions than Embden had expected. Gerhard, however, considered the systems too complex to allow mechanistic understanding; this research was "too much physiology, too little chemistry. The systems got too complicated for my liking. I came to Embden mainly because I was tremendously impressed with reading on liver perfusion . . . and I had expected something more on metabolic mechanisms. And I discussed that with some of the instructors who guided me and instructed me about our techniques. But one day Embden told me that he had heard that I wasn't quite satisfied with the project and, certainly, why didn't I tell him that, because the first condition for successful work must be that the investigator be very much interested in the problem. . . . So Embden told me I should think of something which I would rather do. I shouldn't rush into it, certainly; he would much prefer to have me work on something which I personally liked."

At that time Emben held journal clubs at his home. In some of these journal club meetings, Gerhard reported on some papers of Richard Willstätter on enzyme purification. Gerhard also reported on a 1926 meeting of the German Society of Physicians and Natural Scientists he attended in Düsseldorf, at which a whole morning was devoted to the work of Willstätter (who had resigned from his faculty position by this time) and his associates. After Gerhard's second journal club seminar, Embden said to him, "Well I see you are very much interested in enzyme work. Personally, I'm not so much interested in enzymes—I'm interested in the cell; but why don't you—we are just working on ammonia formation in muscle—why don't you do some enzyme work on ammonia formation from adenylic acid?"

The seed for this new project had been planted in 1914 when, during the fractionation that first yielded lactacidogen from muscle juice, Embden also found a nitrogen- and phosphorus-containing substance of interest.[12] He determined, at that time, that it contained adenine, a pentose and phosphate—but it did not have lactacidogen activity, and he did not pursue it further then. Years later, as reported in 1927, he purified and crystallized the substance and characterized its structure as a nucleotide, adenylic acid (also known as adenylate or adenosine monophosphate; see Figure 4.1).[13]

"Nucleotide" is a general term for a structure in which a nitrogen-containing base (in this case, adenine) is attached to a 5-carbon sugar (ribose or deoxyribose), with a phosphate group linked as an ester to one of the hydroxyl group

---

[12]G. Embden and F. Laquer, "Über die Chemie des Lactacidogens, 1. Mitteilung. Isolierungsversuche," *Zeitschrift für physiologische Chemie* 93(1914): 94–123.

[13]G. Embden and M. Zimmermann, "Über die Bedeutung der Adenylsäure für die Muskelfunktion. 1. Das Vorkommen der Adenylsäure in der Skelettmuskulatur," *Zeitschrift für physiologische Chemie* 167(1927): 137–40.

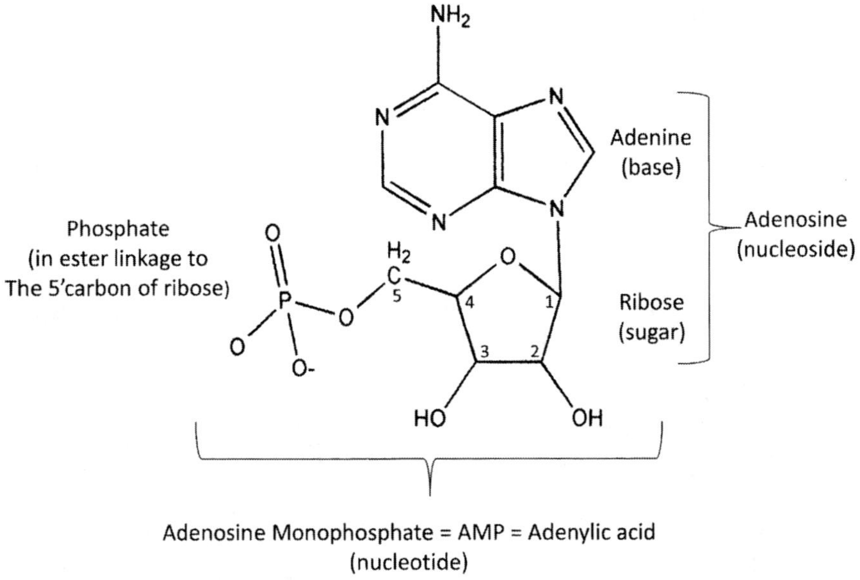

Figure 4.1   Structure of the nucleotide adenosine monophosphate.

oxygen atoms of the sugar. Without the phosphate group, the residual base-sugar structure is a "nucleoside" (for example, adenosine). Nucleotides of the purine bases (adenine and guanine) and the pyrimidine bases (uracil, cytosine and thymine), with either ribose or deoxyribose as the sugar, are the units from which the polymerized nucleic acids, RNA (ribonucleic acid) and DNA (deoxyribonucleic acid), are built. Single nucleotides that are not polymerized also play important roles in physiological chemistry and cell function. Nucleotides had been isolated from DNA more than a decade earlier. Deoxyguanylic acid (guanine–deoxyribose–phosphate), in fact, had been discovered by Siegfried Thannhauser, a brilliant investigator of metabolism and metabolic disease and a man who would eventually play a crucially important role in the development of Gerhard's scientific career. This was pioneering work with nucleotides, decades before the roles of DNA and RNA in the carrying and expression of genetic information became known.

As well as discovering adenylic acid in muscle juice, Embden and colleagues also found that levels of ammonia increased in muscle when it was stimulated to work, and that this increase was reversible;[14] that is, at rest, the ammonia was reincorporated into complex molecules. He wondered whether

---

[14] G. Embden, "Neue Untersuchungen über die Tätigkeitssubstanzen der quergestreiften Muskulatur und den Chemismus der Muskelkontraktion," *Klinische Wochenschrift* 6(1927): 628–31.

release of ammonia from adenylic acid could be involved in the linkage between chemical and mechanical energy. This material and this question provided the basis for Gerhard's habilitation research.

With a stipend from the Notgemeinschaft Deutsche Wissenschaft—Emergency Association of German Science—Gerhard undertook the purification and characterization of the catalytic activity in muscle juice that deaminated adenylic acid, that is, removed the amino group ($-NH_2$), converting it to free ammonia (NH_3). This discovery of the enzyme *adenylate deaminase* was his first major work.

He published his results, as sole author, in the *Zeitschrift für physiolgische Chemie* in 1928.[15] In the article he reported that minced muscle, or the juice pressed out of minced muscle, contained a basal amount of ammonia but the amount increased when he added adenylic acid. That is, the juice contained an enzyme that could remove the amino group from adenylic acid, forming ammonia. The crude juice also acted on adenosine (which, as noted above, has adenine and pentose but no phosphate), but not on free adenine itself. He obtained considerable purification by first doing a preliminary extraction of the minced muscle with neutral saline solution, which dissolved many other proteins, including those that gave red color to muscle, leaving the enzyme activity in the insoluble tissue components. After this preliminary step, the enzyme could then be extracted into solution, from the residual insoluble tissue, with dilute, slightly alkaline sodium bicarbonate, yielding a colorless solution with a high level of enzyme activity.

This purified enzyme was much more active on adenylic acid (the nucleotide) than on adenosine (the nucleoside) and, like the original juice, it did not act on adenine (the free base). Nor did it deaminate a different purine base, guanine, its nucleoside (guanosine) or nucleotide (guanylic acid), or any of 13 different amino acids. The activity on adenosine was much more easily inactivated by mild alkali than the activity on adenylic acid, suggesting there were two distinct enzymes, one of which was really highly selective for the nucleotide. He demonstrated that fact more conclusively by showing that the enzyme acting on adenylic acid could be adsorbed to insoluble alumina particles, whereas the adenosine-directed activity could not; this procedure yielded additional purification. This article provided information on the pH optimum and the kinetics of the reaction catalyzed by the newly defined enzyme, *adenylic acid deaminase*. Describing the kinetics, Gerhard reported values for the rate constant but reported he could not get a true first-order reaction rate unless he used very large amounts of substrate. This published statement came back to haunt him about a year later.

---

[15] G. Schmidt, Über fermentative Desaminierung in Muskel," *Zeitschrift für physiologische Chemie* 179(1928): 243–69.

In the course of studying the specificity of *adenylate deaminase*, Gerhard came upon another important—unplanned—discovery. For his first experiments, he had been preparing the adenylic acid substrate from muscle tissue, according to known purification procedures. This was, however, difficult and expensive. The laboratory could not afford to purchase large animals, so frog muscles were the main source of substrate as well as enzyme, a fact that explains the large number of frogs in the room next to the one Gerhard used as his living space during some of his student years. Muscles were also in demand for many other experiments in the department. Gerhard decided, therefore, to use a much more readily available form of adenylic acid, which one could make by alkali-catalyzed de-polymerization of yeast RNA, which was abundantly available. So Gerhard prepared some yeast adenylic acid in this way but, to his surprise, it was not deaminated by his enzyme, even though the chemical analysis of the yeast and muscle forms of adenylic acid were identical. It was not that the yeast adenylate contained some inhibiting contaminant, because when muscle and yeast adenylate were mixed and exposed to the enzyme, the amount of ammonia formed corresponded to just what would be expected from the amount of muscle adenylate used. The yeast form neither was attacked nor did it interfere with the enzyme action on the muscle adenylic acid.

Embden, perhaps a little suspicious of the quality of Gerhard's yeast adenylic acid, wrote to Professor P. A. Levene, at the Rockefeller Institute in New York, asking him to send an authentic sample prepared in the Levene lab. Levene was one of the leading carbohydrate and nucleic acid chemists and had described a procedure for purification of yeast adenylic acid. The authentic sample from his laboratory did not serve as a substrate for *adenylate deaminase* any more than the yeast adenylate prepared by Gerhard had.

Gerhard reported this difference between yeast and muscle adenylate in his original paper on adenylate deaminase, with the important conclusion that, in spite of identical chemical analysis, these two forms of adenylate must be different substances—isomers—and the likely difference is the point at which the phosphate group is attached to the ribose, the 5-carbon sugar, to form the nucleotide. This was the first identification of these kinds of nucleotide isomers.

In principle, phosphate could esterify any of three different hydroxyl groups in the sugar—the 2', 3' or 5'- hydroxyls (see Figure 4.1). It was known that alkali degradation of RNA cleaves the polymer in a way that releases adenosine-3-phosphate. Until Gerhard's experiments, Embden had no reason to think that the adenylic acid he discovered in muscle was any different. The muscle form, later shown to be adenosine-5-phosphate, was purified directly from the tissue, largely as a spontaneous breakdown product of a nucleotide that has adenine, ribose, and a chain of three phosphates attached to the 5'-OH of the ribose. That molecule, adenosine triphosphate (ATP), which would be discovered in 1929 by Karl Lohmann in Meyerhof's laboratory,[16] became recognized as an important source of chemical energy for muscular contraction and for energy-requiring synthetic reactions.[17] Two of its phosphates are detached readily in solution, leaving the simpler nucleotide adenylic acid (adenosine 5'-monophosphate; AMP).

A year after this first publication, Gerhard and Embden together published a follow-up study with additional data showing definitively that the two isomers were distinct substances.[18] Their conclusions were based on: a slightly higher melting temperature for yeast adenylic acid than for muscle adenylic acid; a marked lowering of the melting point when the two were mixed, below that of either alone; a weaker levorotary optical activity for the muscle isomer; and a much faster rate of phosphate release from the yeast isomer on exposure to acid.

Underlying the interest in deamination of adenylate was the question whether this reaction might be related to the mechanisms of muscle contraction. The amount of ammonia formed, however, was very much less than the amount of lactate, and the link could not be made. Embden and Gerhard did, however, publish an additional paper on deamination of adenylic acid.[19] In this work, they demonstrated that enzymatic deamination of adenylic acid accounted for all of the ammonia formation that occurred spontaneously as isolated muscle tissue fragments broke down by autodigestion. The enzyme was physiologically significant. Still, the amount of ammonia formation from adenylic acid was much too small to account for the energy changes involved in muscle contraction.

Gerhard then turned to different enzymes involved in breakdown of a second purine base, guanine, its nucleoside (guanosine), and its nucleotide (guanylic acid). In fact, this work had been part of his habilitation, but was continued and reported in a preliminary publication of 1931 and a major publication the following year.[20] Having learned the first steps in the metabolic breakdown of adenine-containing nucleosides and nucleotides, he aimed to

---

[16] K. Lohmann, "Über die Pyrophosphatfraktion im Muskel," *Naturwissenschaften* 17(1929): 624–5.

[17] F. Lipmann, "Metabolic Generation and Utilization of Phosphate Bond Energy," *Advances in Enzymology* 1(1941): 99–162.

[18] G. Embden and G. Schmidt, "Über Muskeladenylsäure und Hefeadenylsäure," *Zeitschrift für physiologische Chemie* 181(1929): 130–39.

[19] G. Embden and G. Schmidt, 'Über die Bedeutung der Adenylsäure für die Muskelfunktion: weitere Untersuchungen über die Herkunft des Muskelammoniaks," *Zeitschrift für physiologische Chemie* 186(1930): 205–11.

[20] G. Schmidt, "Über den Abbau des Guaninkerns durch die Fermente der Kaninchenleber," *Klinische Wochenschrift* 10(1931): 165-167; G. Schmidt, "Über den fermentativen Abbau der Guanylsäure in der Kaninchenleber," *Zeitschrift für physiologische Chemie* 185(1932): 208–24.

uncover these processes for all purines in both their free form and in their nucleoside and nucleotide forms.

Before Gerhard's work, it was known that an enzyme, named *guanase*, could catalyze the deamination of free guanine. In this regard, metabolism of guanine differed from that of adenine, which was not susceptible to deamination unless it was bound in a nucleoside or nucleotide structure. Gerhard reported on purification of *guanase* and novel details of its properties—specificity, lability at high temperatures, pH dependence, and kinetics. *Guanase* was shown to be a very selective enzyme, catalyzing deamination only of guanine, with only slight activity on guanosine and none on guanylic acid or any form of other bases, nucleosides, or nucleotides or amino-bearing compounds, such as amino acids.

He found a distinct enzyme that catalyzed deamination of the nucleotide guanylic acid, in a process that was very different from the action of *adenylate deaminase* on adenylic acid. Coincident with the deamination of guanylic acid, there was additional breakdown, with release of free purine and ribose-phosphate. Again, this enzyme was selective for guanylic acid and did not act on adenylic acid or other bases or nucleosides or nucleotides bearing amino groups.

Gerhard then realized that the high degree of selectivity of an enzyme like *guanase* could be put to use in an assay to measure guanine concentrations in relatively small amounts of tissue extracts; together with a student, Ernst Engel, he developed a suitable procedure and applied it to several tissues and organs.[21] As expected, guanine levels were much higher in nucleic acid-rich organs, such as calf thymus, than in liver or muscle. He established the validity of the assay by adding known amounts of guanine to the raw tissue extracts and accounting precisely for the added material in the measured ammonia formation.

He then modified and expanded the procedure into a protocol that could be used to measure guanine and total aminopurine content (including oxypurines) in the tissue extracts; and, from these values, he could calculate adenine

---

[21] G. Schmidt and E. Engel, "Mikrobestimmungen von Purinsubstanzen in Geweben I. Mitteilung: Die Bestimmung des Guanins," *Zeitschrift für physiologische Chemie* 208(1932): 225–36. The analytical procedure involved: preparation of a liquid extract from tissue minced in liquid air and 2% sulfuric acid, reflux heating and condensation in sulfuric acid to completely split free purines from bound compounds in which they occurred, neutralization of the acid, removal of any pre-existing ammonia by vacuum distillation, exposure of the extract to purified guanase and measurement of ammonia formed by that exposure.

content as well.[22] This work marked the beginning of two of Gerhard's career-long interests: the development of improved methods for precise measurement of biologically important substances, whether nucleotides, nucleic acids, or phospholipids; and the biochemical properties and importance of phosphate esters.

He began research in a different direction as well. Taking note of important work done by Karl Lohmann on phosphate esters, in the laboratory of Otto Meyerhof, Gerhard began to isolate peptide-phosphate breakdown products of the phosphoprotein casein. He did not have time to publish results before he had to leave Germany in 1933. The second paper on purine assays and one on dipeptide-phosphate were submitted from Naples.

---

[22]Schmidt, "Mikrobestimmungen von Purinsubstanzen in Geweben. 2. Mitteilung. Die Bestimmung des Guanins, des Adenins und der Oxypurine," *Zeitschrift für physiologische Chemie* 219(1933): 191–206. The first steps were: precipitation of proteins and other large molecules with tungstic acid, acid hydrolysis of purine-containing molecules to release purine bases, and extraction of all the purines into solution with hydrochloric acid. A portion of this extract was used for measurement of total purine content. All purines were precipitated with copper sulfate and sodium bisulfite; total nitrogen content of the precipitate was measured with the Kjeldahl method and the value was converted to its purine equivalent. A second portion of the hydrochloric acid extract from the tungstic acid precipitate was used for total aminopurine and guanine analysis. With a known procedure, all aminopurines, together with most of the oxypurines, were selectively and quantitatively precipitated with ammoniacal silver solution. The precipitate was washed, purines redissolved in HCl (hydrochloric acid), and insoluble silver chloride removed by centrifugation and filtration. The purine solution was neutralized, existing ammonia distilled away, and the nitrous acid was used to deaminate all aminopurines; resulting ammonia was measured and total aminopurines calculated from this value. A second, smaller portion of the original extract was used for specific measurement of guanine with the enzyme guanase. The difference between total amino purine and guanine gave the value for adenine. The difference between total purine and aminopurine values gave the value for oxypurines.

Five

Great Days, but Looming Danger

A lthough dangerous political winds were gathering in the late 1920s, conditions during the years between 1925 and 1929 still seemed relatively stable, economically sound, and open to free expression and creativity in many fields. Academic budgets were not high, but laboratories were well equipped and, even in times of financial difficulty, giant firms in the chemical industry helped to keep academic laboratories active. There were several such firms near Frankfurt. One kind of help given was provision of materials, such as liquid nitrogen. Almost daily, someone from Embden's lab could take a container to the company, have it filled with liquid nitrogen, and return it to the university, where it was used in storage of tissues and biochemical products.

Laboratories had glassware of high quality, notably from Jena, although items such as individually fitted ground-glass stoppers and precisely calibrated pipettes were precious. Electrical centrifuges were available, as well as very large filtration funnels and a giant mortar and pestle. Indeed, German scientists who fled to the United States during the 1930s often found laboratories in their new country much less well equipped.

Economically, the situation became manageable. Even during inflation, academic salaries had been adjusted accordingly. In 1925, during his internship at the hospital in Stuttgart, Gerhard lived at home; and, as an intern, he had free meals at the hospital. When he returned to Embden's lab in Frankfurt for the last three months of the internship, Embden arranged a Rockefeller scholarship for his support. Life under these conditions was quite agreeable for a young person. Nobody had any reserves, but the salaries were adequate. Having experienced the inflation, young people at that time, including Gerhard, were inclined to enjoy life as long as it was possible. "We went to all concerts in Frankfurt." The years after the inflation were very stimulating years, although one enjoyed them without much confidence in the future.

Work in the lab was also enjoyable, though there were some tense moments. In preparing crystals of pure adenylic acid for the studies on *adenylic acid deaminase*, for example, Gerhard did something that tested Embden's good humor. It was important that the adenylic acid crystals be dry, and the last traces of water were removed by distillation, with a pistol-shaped glass condenser–flask combination, named the Abderhalden Pistol after its inventor.[1] The crystals were placed in a horizontal barrel connected to a vacuum source, and this tube was within a larger glass barrel connected to a secondary flask, which contained toluene and was connected to a condenser. In this apparatus, the toluene could be boiled to produce warm vapors of known temperature surrounding the inner barrel, providing heat that helped to drive residual water out of the crystal preparation. The toluene-containing flask was heated with a burner. Because it was expensive and very difficult to have precisely fitted glass-to-glass connections to form a fine seal between the flask and the condenser, a cork stopper was used instead. "Since the crystal was not quite dry, there was just a drop of water sitting in that inner retort, which is heated. The temperature would never go above 100°C, whereas it should go to 120°C. So I turned the flame somewhat higher, with the result that on one of the joints the toluene vapors caught fire, and as you know toluene is famous for being used for making kymographs black (i.e., forming a soot-blackened layer on the surface of a drum used for recording muscle contractions with a stylus that scratches a line in the soot layer). After three minutes you couldn't see any more in the laboratory. There was an older technician, Fraulein Fleck, across on the other side of the bench, and she just disappeared. It was completely like one of those stories in "Max and Moritz"—you know, by Busch.[2] It was completely covered with soot." What is more, Gerhard's own lab was being painted at the time, and it was Embden's freshly painted lab that he had covered with soot. Embden was away at the time. Gerhard was apprehensive because there were precious substances stored in glass jars, the surfaces of which were all blackened. However, the lab was eventually cleaned by the diener, and Embden seems to have taken the event in stride. Gerhard's career continued.

A second test of Embden's graciousness came after the papers on *adenylate deaminase* were published. As noted previously, although the first paper established very important discoveries and became well known, there was one statement that led to some embarrassment. The paper said that, except

---

[1] A glass apparatus for vacuum drying of crystals, designed to remove the last traces of water, was described by Carl Brahm and J. Wentzel, a student and lecturer, respectively, under Emil Abderhalden, in the latter's *Handbook of Biochemical Methods*, 1906. It consists of "a cylindrical double-walled vacuum chamber (the barrel) sitting between a flask containing solvent below and a reflux condenser above" and a flask with a bent neck (the handle) containing a drying agent. Abderhalden, a medical school graduate, worked with Emil Fischer, became Director of the Kaiser Wilhelm Institute in Halle; Andrea Sella, "Classic Kit: Abderhalden's Drying Pistol," *Chemistry World*. Royal Society of Chemistry, retrieved from http://www.rsc.org/chemistryworld/ February 2009.

[2] *Max und Moritz—Eine Bubengeschichte in sieben Streichen* ("Max and Moritz—A Story of Seven Boyish Pranks") by Wilhelm Busch, published in 1865, an illustrated children's book composed in rhymed couplets in German.

at high substrate concentration, the data did not fit a theoretical curve for generating a kinetic constant for the reaction. After its publication, Gerhard and Embden received a letter from Karl Myrbäck, an eminent biochemist and enzymologist at the University of Stockholm, saying that he had read the paper with great interest, and it agreed with some experiments that he and his colleagues did with muscle adenylic acid. "But he wanted to call our attention to the fact that our data were correct, but these are the data of a very good reaction constant. We only forgot to take the logarithm. And, since I did all these calculations, Embden was not too pleased and we had, of course, to send in the corrections to Hoppe-Seyler's *Zeitschrift für physiologische Chemie*."[3]

A FIRST VISIT TO AMERICA

A great event for Gerhard, the International Congress of Physiology, was held in Boston in August 1929 (there were no separate biochemistry societies or meetings at that time). This was just after publication of the papers on *adenylic acid deaminase*. Embden obtained some funds from industry to allow Gerhard to attend the Congress. He sailed from Le Havre to New York, along with about 700 European physiologists and biochemists, including great scientists, such as Walter M. Fletcher and Joseph Barcroft, from Cambridge University, and Archibald V. Hill and Henry H. Dale from London. "These were all marvelous sportsman-like giants . . . English aristocracy and intellectual aristocracy. I was very much flattered when, one morning, I came late for breakfast to the dining room and there was one seat at Fletcher's table, and he waved to me so I had breakfast at the same table as Fletcher." During the voyage, there was a memorable tug of war (a real rope-tugging battle) between an English team and some German representatives, among them Emil Abderhalden and Franz Knoop. In a blow to German pride, the English team won. On the other hand, many German scientists, including Embden, had traveled on a different boat, the *Bremen*, one of the newly built products of a recovered German ship-building industry and a source of pride for the nation; it had been launched in 1928. Perhaps it carried some stronger Germans than those on Gerhard's vessel.

Among Gerhard's co-passengers was a professor of physiology from Innsbruck. "He had a very pretty daughter who impressed me very much. But unfortunately I had as competitor—a very elegant Hungarian who was no one

---

[3] G. Embden and G. Schmidt, "Berichtigung," *Zeitschrift für physiologische Chemie* 197(1931): 191–92.

else but Szent-Györgyi[4] in his best years, who had the polished manners of a Hungarian aristocrat; and he won out. I mean I couldn't even get a dance with her. Szent-Györgyi stole all the waltzes."

At the Congress in Boston Gerhard met, for the first time, Karl Lohmann, who had just discovered adenosine triphosphate (ATP), a critically important source of energy for muscle contraction and for many physiological chemical reactions. Lohmann was from the laboratory of Otto Meyerhof, the scientific camp sometimes in contention with that of Embden. Gerhard recalls, "Somebody introduced me, and immediately —and this was very characteristic for Lohmann—the first sentence after we were introduced was 'I am very greatly honored to meet you, dear colleague Schmidt. I have read with interest your recent publications. You seem to be already in that age where you only publish corrections of your former papers.'"

ATTAINING FACULTY STATUS

On his return to Germany from the 1929 Congress of Physiology, although Gerhard had completed the research for his habilitation, Embden did not have a position available for him, but recommended him to Dr. Bernhard Fischer-Wasels, chairman of the Department of Pathology at the same University of Frankfurt. On Gerhard's behalf, Embden added a condition to his recommendation. He told Fischer, "Please, Schmidt would like to work with enzymes. Please let him work with enzymes. Don't interfere with his work. He could certainly take responsibilities, supervise analysis, and things like that, but don't interfere with his own plans." And Fischer was completely agreeable to this; he was an extremely upright and fair-minded person in this matter, as he was in general.

Fischer-Wasels recognized the importance of the growing field of biochemistry and was interested in having a laboratory of biochemical pathology within his department. He followed currents in biochemical research, without being an expert himself, a condition that led him into some fruitless research. For example, he was aware of Otto Warburg's great contributions, including the

---

[4] Albert Szent Györgi (1893–1986), a Hungarian scientist. He received his PhD in 1927 from the University of Cambridge, UK, where he isolated a hexuronic acid, which eventually became known as ascorbic acid or Vitamin C. He won the 1937 Nobel Prize in Physiology or Medicine for studies of biological oxidation and the role of ascorbic acid. After World War II, he developed his career in the United States, turning to mechanisms of muscle contraction; *Nobel Lectures, Physiology or Medicine 1922–1941* (Amsterdam: Elsevier, 1965). Online biography retrieved October 15, 2012, from "Albert Szent-Györgyi—Biographical," http:// www.nobelprize.org/ nobel_prizes/medicine/laureates/1937/szent-gyorgyi-bio.html.

discovery of unexpected high rates of glycolysis in cancer cells even under conditions of active aerobic oxidative respiration and his theory that high glycolysis activity even in the presence of oxygen was connected to the pathogenesis of the cancer.[5] Fischer-Wasels concluded that one should treat cancer patients by exposing their tissues to high levels of oxygen to force the cancer cells into oxidative respiration rather than glycolysis: "If the cancer cell doesn't want to breathe, one has to force it to breathe." He persuaded some wealthy citizens of Frankfurt to provide funds for trying out this idea on hopeless cancer patients. Barracks were established in the hospital, with a fair number of patients, and these patients were put in a kind of oxygen tent and were exposed to pure oxygen for an hour once or twice a day.

Gustav Embden was, perhaps, Fischer-Wasel's best friend at the university. Embden knew that Fischer's approach was very unlikely to work; but it was very difficult to dissuade a clinician from trying out his passionately held hypothesis. Being more aware of basic physiology than Fischer was, Embden was at least able to ensure that 5% carbon dioxide was added to the oxygen so that patients' breathing would continue to be stimulated. As sometimes happens in such trials, there were a few patients who apparently responded to this treatment. In the end, it had to be abandoned; but this interest was one of the reasons why Fischer wanted to establish a chemical laboratory. Gerhard became head of that biochemistry laboratory.

Around that time a second theory that stimulated Fischer's interest in chemistry was proposed by a Belgian surgeon, René Reding, in a book called *Cancer and the Cancer's Territory*,[6] in which he proclaimed that cancer was invariably accompanied by and probably caused by abnormal blood pH. Although Gerhard's lab had a good electrometric pH meter, it was cumbersome for use with many samples. Reding was applying a colorimetric pH method, designed by Baird Hastings especially for measuring blood pH. Fischer-Wasels sent Gerhard to Reding's clinic in Belgium for three weeks, expenses paid, to learn Reding's technique for measuring tissue pH so he could come back and show the technicians in Frankfurt how to do it.

Gerhard joined the Department of Pathology with the rank of assistant. He was not yet eligible for full faculty rank. He achieved that goal when his habilitation thesis, which included the work on guanine nucleotide breakdown and the enzyme *guanase*, was accepted in 1931. At that time he attained the

---

[5] O. Warburg, K. Posener, and E. Negelein, "Über den Stoffwechsel der Tumoren," *Biochemische Zeitschrift* 152(1924): 319–44; O. Warburg, "On the Origin of Cancer Cells," *Science* 123(1956): 309–14.

[6] René Reding, *Le Terrain cancéreux et cancérisable. Physiologie pathologique du cancer. Action biologique des radiations* (Paris: Masson, 1932).

faculty title of privatdozent, the first step in the academic ladder. He worked independently, on his own research projects, but also provided some service for the Pathology Department.

DARKENING POLITICAL CLOUDS

The bohemian enjoyment of life did not last long after Gerhard's return from the International Physiological Congress of August 1929. Just two months later came the great stock market crash. The international financial crisis profoundly affected Germany's economy, which had depended on international loans, particularly from American sources. Unemployment and the inability of the Weimar government to fulfill social support obligations led to civil unrest. The democratic governing system, the Weimar Republic, was becoming increasingly unstable.

At the same time, it was becoming evident that the Nazi Party was becoming an increasingly significant force, both in violent episodes in the streets and in electoral politics.[7] In the previous year, in the Reichstag elections of May 20, 1928, the Nazis received 0.8 million votes, 2.6% of the total, gaining just 12 of 429 seats. The government formed after the 1928 elections, led by the Social Democrat Hermann Müller, lasted only until March 27, 1930, when President Hindenberg dismissed Müller and appointed Heinrich Brüning, of the Center Party, a mainly Catholic party, as Reich Chancellor. New Reichstag elections were called for September 14, 1930. This time the Nazi Party increased its vote to 6.4 million, 18.3% of the total, gaining 107 of the increased number of 577 seats.[8]

In the four years between 1929 and 1933, one heard more and more about persons among one's immediate acquaintances becoming part of the Nazi movement; one of the young caretakers in Embden's lab, who had always been considered a socialist, became a member of the Party. In a time of economic crisis, unemployed workers saw some hope in the Party's proclamations; but even middle-class Germans, who had not supported the Nazi Party earlier, found some identification with its nationalist aspects. This shift was partly an enduring response to the terms of the Versailles treaty, with its burden of reparations, loss of lands in the east and west and diminution of Germany's power and international position. Gerhard recalls that although Germany maintained its sure sense of superiority in achievements in science, arts, and industrialization, it had a sort of self-pity in relationship to international politics. Such feelings affected even people like Embden, who had spent

---

[7] R. J. Evans, *The Coming of the Third Reich* (New York: Penguin Books, 2004), 247–65.
[8] Evans, *The Coming of the Third Reich*, 259–61.

very important formative years at the University of Strassburg/Strasbourg, with Hofmeister; he was always very homesick for that city. He belonged to an association of former Alsatians, although he was originally from Hamburg. Embden, of course, never came near supporting Nazi politics; but for him, the loss of Alsace-Lorraine to France was an excruciating humiliation in foreign politics. As nationalistic feelings, heightened by territorial and economic losses, moved many people toward the feeling of self-pity in external affairs that Gerhard described, he noted that there came a time when many middle-class people could say, "Of course, I should be against National Socialists coming to power; however, don't you admit that there is very much idealism in their movement?"

Brüning remained in office after the September 1930 election, but his unpopular response to the financial crisis—lowering wages and prices—led to collapse of his government in May 1932.[9] During his tenure, steps were taken toward authoritarian government, as rule was increasingly imposed under presidential emergency powers. General Paul von Hindenberg, who had just defeated Hitler in being re-elected as Reich President in April 1932, appointed the conservative Center Party member Franz von Papen as Reich Chancellor on June 1. Papen and his cabinet had no parliamentary support and the Reichstag was dissolved a few days later.

In new elections, on July 31, 1932, the Nazi Party increased its vote to 37.4% of the total (230 of 608 seats), the highest of any party, though still not an outright majority of votes.[10] Papen, anti-Nazi, established a minority government of conservatives, including a reactionary Minister of the Interior, Franz Bracht. In spite of its authoritarian bent, it had little support, and its actions contributed to public scorn. Gerhard recalled that, in the midst of economic disaster and political chaos, the first act of the Papen cabinet was to pass the "Zwickelerlass," a law enacted August 18, governing the design of men's bathing suits in order to protect the morals of the people. Tabloid newspapers had headlines like: Jeder Deutsche Badehosen mit ein Zwickelbar (Every German Swimsuit has a Gusset Bar). "You can't imagine. This was the one and only occasion where all parties, including the National Socialists and Social Democrats, Democrats, laughed together—the first and last time, because for the National Socialists, of course, this was water on their mill to overcome this Conservative."[11]

By September 12, the Papen government was soundly defeated by a no-confidence motion raised by Herman Göring in the Reichstag, and new

[9] Evans, *The Coming of the Third Reich*, 271–88.

[10] Evans, *The Coming of the Third Reich*, 293.

[11] "Kulturkampf um Badekleidung Zwickel im Schritt," *Der Spiegel* August 1932. Retrieved October 17, 2012, from *Der Spiegel* online at http://einestages.spiegel.de/s/tb/25353/zwickelerlass-gesetz-zur-bademode-1932.html.

Reichstag elections were called for November 6. This time the Nazi Party vote declined, from 13.7 million to 11.7 million (from 230 seats to 196 seats), whereas Nationalists and Communists—both of which were enemies of the Weimar Republic and its parliamentary system—gained.[12] Though Nazi support seemed to have peaked and fallen, anti-Nazi forces were deeply divided among themselves; even on the left, Communists considered Social Democrats to be their main enemy. Unable to govern, Papen resigned and General Kurt von Schleicher was appointed Chancellor on December 3. By now, the Reichstag was marginalized and presidential decree became even more central. Hitler refused Schleicher's entreaties to join a coalition cabinet. Only as Chancellor would he form a government. Hindenberg and Papen sought to control and limit the Nazis by bringing them into the government. Frantic negotiations, along with military and conservative nonmilitary forces, led Hindenberg to offer the Chancellorship to Hitler on January 30, 1933, with certain strings attached and with a majority of non-Nazi cabinet members. The attached strings were easily broken. Hitler and the Nazi Party soon took over both in the streets and as the ruling power.

SCIENTIFIC LIFE UNDER THE CLOUDS

In this period of turmoil in the political world, research continued at the university. With his work on purine breakdown leading to an interest in embryo chemistry, Gerhard had begun work with chicken egg embryos. Even though the egg is large, he determined that the yolk has practically no purines, so he wanted to study purine chemistry in eggs that were much smaller, where the yolk was not present in such overwhelming excess. His choice was sea urchin eggs, which were small, had little yolk, and had "this beautiful mode of total cell division until the morula stage." Early in the summer of 1932 he traveled to Helgoland, near Hamburg. Helgoland was a small island, a red bank of sand with white, steep rocks and, on top the island, a very green plateau—"one of the most beautiful spots in the North Sea"—which Germany purchased from England near the turn of the twentieth century. There was a small marine biology station, where one could get sea urchins.

The visit to Helgoland was a pleasant experience. Because, under the terms of its purchase, it was a free port, one could afford luxurious English or Egyptian cigarettes, lobsters, shrimp, and the best French wines. Although the experience was pleasant, the equipment at the marine biology station was primitive, not suitable for significant experimental work.

---

[12] Evans, *The Coming of the Third Reich*, 298–301.

Later in the summer, Gerhard and Embden traveled to Rome to attend the fourteenth International Congress of Physiology August 29–September 3, 1932. Gerhard had become increasingly interested in the varieties of phosphate linkages in organic compounds, such as nucleotides, and had turned to examining phosphoproteins. At the Congress he delivered a talk in which he presented some unpublished results on the novel isolation of a phosphodi-peptide from the phosphoprotein casein. He had determined, by elemental analysis and a variety of tests, some of which excluded various other amino acids, that his compound contained glutamic acid, serine, and phosphate. In the next year he would wish that he had actually published that result in a journal.

It was the first time he experienced Italy. "Also it was, for quite a number of us, the first time we saw, first-hand, life in a fascist country. The conference was opened by Mussolini. Participants had to pass seven cordons of carbonieri, before they entered and, at each cordon, to present the invitation before being admitted." When everyone was seated, anticipating representatives of the Italian government, Mussolini marched in, with a brisk step, followed by his cabinet. Participants were urged to attend a special exhibition on fascism: mainly pictures of local party bosses and their exploits."

Italian newspapers provided the strangest experience: "They were very short, perhaps only six pages, but the front page almost every day contained a photograph of Mussolini on a white horse in some official celebration or inauguration. By far the largest part of the text of the newspaper was advertising of the propaganda for fascism. At that time it was still comical for somebody who came from a country like Germany, from the Weimar Republic, particularly from Frankfurt, where the newspaper was the most objective, most well-informed, best newspaper in Germany, the *Frankfurter Zeitung*, where we were accustomed to this type of newspaper. . . . Oh yeah, the Nazis published the *Völkischer Beobachter*. I never. . . . It was too low for me to look at . . . and *Der Stürmer*, the anti-Semitic newspaper of Julius Streicher was too dirty to touch, repulsive in its propaganda. There were certainly newspapers of the Conservatives, nationalistic, certainly polemical; but they used relatively decent language. This actual experience of fascism made me less pessimistic than I was before, because I just couldn't picture something like that taking place in the near future in Germany."

Gerhard perceived, therefore, in late 1932, that basic institutions, such as the press, were still able to operate freely in Germany and that one could express one's personal opinion, and that universities still seemed to work under the motto from imperial times: "die Wissenschaft und der Lerher ist frie"—"Science and its teaching are free." Although pressures and restrictions were being felt more and more, at that time "Nobody imagined the degree to which the expression of dissenting opinions would be suppressed."

Apart from the exposure to fascism, Gerhard found Italy to be magnificent, the "dream of every German romanticist," particularly in its art and landscape. After the Congress, he spent some time at a marine biology research station in Naples, collecting sea urchin eggs, useful materials for his research. This visit was the beginning of Gerhard's enduring love for Italy and a recurrent desire to be there. Hoping to achieve more with sea urchin eggs than he had been able to accomplish in Helgoland, he applied for rental of bench space at the Naples marine biology lab for the following year, 1933. Without knowing how significant this application and timing would be for his future, he applied for the space between April 1 to April 30 because that would be the season of sea urchin mating.

The Marine Biology Station at Naples was originally founded in 1870 as a private enterprise of the Dohrn family, but was taken over by the state during the First World War.[13] After the war, the station was sustained by renting lab space to scientists, many from other European countries or America. The collaborating countries contributed funds for maintenance. In order to rent space, one had to apply to his or her home country government, which would appoint one to the space without cost but would not provide other funds for salary or supplies. It was still a few months before Hitler's chancellorship when Gerhard applied in 1932. He had no difficulty in having the German government approve his application.

Back at home, later in 1932, Gerhard met Karl Lohmann again, at a grand meeting of the Kaiser Wilhelm Gesellschaft in Heidelberg—the last such meeting there before Hitler came to power. At the meeting were Meyerhof, Warburg, Wieland, Neuberg,—the best of biochemists.[14] As Gerhard's friend, Werner Lipschitz, the chairman of pharmacology at Frankfurt, who was there, put it, "It was an audience of kings." Because the meeting was in Meyerhof's institute, however, Embden did not go even though he had been invited. Gerhard had seen the invitation (with wonderful paper and silk paper inserts, it looked as if it were for a wedding) on the desk of Embden's secretary, and it seemed "terribly interesting." Without asking Embden, Gerhard took a day off and took the train for the one-hour ride to Heidelberg for the chance to hear Otto Warburg and Carl Neuberg. When asked at the registration desk for his invitation he said, "I just came. Couldn't I listen to the lectures here?" After a moment, a door opened and out came Karl Lohmann, who "solidified like Lot's wife," and said "You are here? What are you doing here?" "For heaven's sake, I just saw this invitation. It looked very interesting and I

---

[13] A history of the Stazione Zoologica. Anton Dohrn is presented on its website, accessed October 17, 2012, at http://www.szn.it/SZNWeb/showpage/107?_languageId_=2.

[14] Otto Meyerhof, Otto Warburg, and Heinrich Wieland had all received a Nobel Prize. Carl Neuberg had been nominated four times for the award in physiology or medicine.

thought I would come." "Okay." So Lohmann went back to Meyerhof's office, gave Gerhard an invitation, and he was able to register. The experience left a lasting impression.

In that autumn of 1932, Gerhard was with Embden just as Embden was realizing a field-transforming insight into the chemical pathway of glycolysis.[15] It was generally accepted by then that glycolysis was a source of energy for muscle contraction, and that the form of energy coupling to contraction was not simply putting heat into the system as in many machines. Otto Meyerhof's laboratory had made many contributions to understanding the thermodynamics of muscle contraction and thought for a while that generation of lactic acid was most closely coupled to it. Embden disputed the concept of direct coupling of lactic acid formation and muscle contraction and, in fact, Emil Lenhartz, a colleague in Embden's lab, and Ejnar Lundsgaard had obtained data showing that lactic acid formation continued after cessation of tetanic muscle contraction.[16]

In the late 1920s, phosphate-containing compounds that generated abundant energy on hydrolysis of the phosphate bond—usually a phosphoric acid anhydride rather than a phosphate ester—were discovered in muscle and their hydrolysis was considered to be even more closely linked to contraction than was lactic acid production. Karl Lohmann, in Meyerhof's lab, discovered and showed the potential importance of creatine-phosphate and then adenosine triphosphate for this coupling function.[17] Hydrolysis of ATP has since been considered as the most directly coupled energy-providing chemical reaction, and creatine-phosphate as the reservoir of high energy phosphate.

Questions remained about how the high-energy phosphate compounds were generated, and it was concluded that glycolysis must play some role. It was important to learn the whole chemical strategy of the glycolytic process, which was being studied in parallel in two kinds of systems: yeast fermentation of sugar that produces ethanol, and animal tissue glycolysis that produces lactate. Pyruvate was a near-end product in both cases and could be converted to ethanol in yeast or lactate in muscle. It was known since 1906 that, in yeast,

[15] For a comprehensive history of how the process of glycolysis was unraveled, see M. Florkin and E. H. Stotz, *Comprehensive Biochemistry, Volume 31. A History of Biochemistry Part III. History of the Sources of Free Energy in Organisms* (Amsterdam: Elsevier, 1975). A useful history is found in A. J. Barnett "A History of Research on Yeasts 5: The Fermentation Pathway," *Yeast* 20(2003): 509–43.

[16] G. Embden and E. Lenharts, Der zeitliche Verlauf der Milchsäurebildung bei der Muskelkontraktion. II. Mitteilung," *Zeitschrift für physiologische Chemie* 176(1928): 231–48; G. Embden, "New Investigations on the Chemistry of Muscle Contraction," *Klinische Wochenschrift* 9 (1930): 1337–40. E. Lundsgaard, "Unterschungen über Muskelkontraktionen ohne Michsäurebildung," *Biochemische Zeitschrift* 217 (1930): 162–77.

[17] O. F. Meyerhof and K. Lohmann, "Über eine neue aminophosphorsaure," *Naturwissenschaften* 16(1928): 47; O. Meyerhof and K. Lohmann, "Uber den Ursprung der Kontraktionswarme," *Naturwissenschaften*,15(1927): 670; "Über energetische Wechselbeziehungen zwischen dem Umsatz der Phosphorsaureester im Muskelextrakt," *Biochemische Zeitschrift*, 253(1932): 431.

inorganic phosphate is required to sustain fermentation; and sugar phosphate esters were known to accumulate when yeast fermentation or muscle glycolysis, which were eventually recognized to be basically the same process, was blocked.[18]

One conceptual problem that had to be solved was the fact that these processes converted a six-carbon sugar like glucose, which does not contain phosphate, to three-carbon products like pyruvate, none of which contained phosphate—and yet phosphate was required and sugar-phosphate compounds could be used as substrates in the process. In addition to phosphate esters of glucose and fructose, some phosphate esters of three-carbon sugars (trioses) had been identified, but it was not clear how or whether they were part of the chemical pathway. In fact, Carl Neuberg had postulated a pathway in which they were not involved; rather, he proposed that nonphosphorylated methylglyoxylate was a key intermediate.[19] Clearly, it was a complex process. A member of Meyerhof's lab published an article claiming to have discovered and solubilized "the enzyme of glycolysis,"[20] a claim at which Embden scoffed because he knew that much more than a single enzyme was involved. In some exchanges, members of Embden's lab and Meyerhof's lab found themselves in confrontation even though Embden and Meyerhof held each other in deep respect.[21]

Embden's earlier work had shown that hexose-diphosphate could be used in muscle glycolysis but he was, frustratingly, not able to isolate the compound from actively glycolyzing muscle.[22] A landmark experiment had been reported by Ragnar Nilsson, working with Hans von Euler in Stockholm, but its importance was not immediately recognized. Nilsson discovered that, in muscle poisoned with fluoride, there was an accumulation of the three-carbon compound phosphoglyceric acid, a clue to the direct involvement of such compounds in the pathway.[23] Even in discussing this result with Nilsson, members of the Meyerhof laboratory did not consider triose-phosphates as

---

[18] A. Harden and W. J. Young, "The Alcoholic Fermentation of Yeast-Juice," *Proceedings of the Royal Society of London B* 77(1906): 405–20. A. Harden and W.J. Young, "The Alcoholic Ferment of Yeast-Juice. Part III.—The Function of Phosphates in the Fermentation of Glucose by Yeast-Juice," *Proceedings of the Royal Society of London B* 80(1908): 299–311; A. Harden, *Alcoholic Fermentation* (London: Longmans, Green,1914).

[19] M. Florkin and E. H. Stotz, *Comprehensive Biochemistry, Vol 31 A History of Biochemistry Part III. History of the Identification of the Sources of Free Energy in Organisms* (Amsterdam: Elsevier, 1975), 73–76.

[20] Possibly referring to O. Meyerhof and K. Meyer, "The Purification of the Lactic-Acid-Forming Enzyme of Muscle," *Proceedings of the Physiological Society, Journal of Physiology*, 64(1927): xvi; or O. Meyerhof, K. Lohmann, and K. Meyer, "Über das Koferment der Milchsäure-bildung im Muskel," *Biochemische Zeitschrift*, 237(1931): 437.

[21] D. Nachmahnson, *German-Jewish Pioneers in Science, 1900–1933: Highlights in Atomic Physics, Chemistry, and Biochemistry* (Berlin: Springer-Verlag, 1979), 333–36.

[22] G. Embden and M. Zimmerman, "Über die Chemie des Lactacidogens. V.," *Zeitschrift für physiologische Chemie* 167(1927): 114–36.

[23] R. Nilsson, "Studien über den enzymatischen Kohlenhydratabbau,"*Arkiv för Kemi Mineralogi Och Geologi*, 10A(1930): 1–140.

key intermediates in glycolysis.[24] It was Embden's insight that showed how they were involved. Gerhard recalled that Embden always considered the education of his trainees to be of prime importance, and as part of that education he had each one of them isolate and crystallize an organic compound. He assigned one trainee, Hans Joachim Deuticke, to a very complicated and difficult chemical purification procedure aimed at obtaining crystals of a sugar phosphate from muscle. It required working with large amounts of material and large-scale precipitation and filtration. At the end Deuticke obtained some crystals and had them analyzed. They were phosphoglyceric acid.

Within a day of that result, Gerhard was walking in the hallway with Embden, and Embden told him that, with the isolation and crystallization of phosphoglyceric acid, he now knew the whole glycolytic pathway. He proposed that the six-carbon fructose-1,6-diphosphate was cleaved into two triose phosphates, glyceraldehyde-phosphate and dihydroxyacetone phosphate, and that these two compounds underwent a known kind of rearrangement, a Cannizzaro reaction, wherein dihydroxyacetone phosphate would be reduced to glyceraldehye-phosphate and the glyceraldehye-phosphate would be oxidized to phosphoglyceric acid; furthermore, he noted that glyceric acid is simply a hydrated form of pyruvate, so he envisioned a full pathway from hexose-bisphosphate to lactate (two additional intermediate steps, between phosphoglycerate and lactate formation, were discovered later). His test of the hypothesis was performed within days; indeed, when phosphoglycerate was added to a muscle extract undergoing glycolysis, it served as a substrate for the increased production of pyruvate and lactate. He submitted a publication to *Klinische Wochenschrift* and another to *Biochemische Zeitschrift*; they appeared early in 1933.[25] Everyone in the field, including Meyerhof, recognized Embden's breakthrough achievement.[26] The glycolytic pathway, in fact, came to be known as the Embden–Meyerhof or Embden–Meyerhof–Parnas pathway, the latter form recognizing the contributions of Jacob Parnas, a Polish biochemist who had also trained with Hofmeister in Strassburg, had spent some time with Meyerhof, and established his own strong group at L'viv University.

---

[24] F. Lipmann, "Reminiscences of Embden's Formulation of the Embden-Meyerhof Cycle," *Molecular and Cellular Biochemistry*, 6(1975): 171–75.

[25] G. Embden, H. J. Deuticke, and G. Kraft, "Über das Vorkommen einer optisch aktiven Phosphoglycerinsäure bei der Glykolyse in der Muskulatur," *Zeitschrift für physiologische Chemie* 230(1933): 12–28; G. Embden, H. J. Deuticke, and G. Kraft, "Über die intermediaren Vorgänge bei der Glykolyse in der Muskulatur," *Klinische Wochenschrift* 12(1933): 213–15.

[26] O. Meyerhof, "Intermediate Products and the Last Stages of Carbohydrate Breakdown in the Metabolism of Muscle and in Alcoholic Fermentation," *Nature* 132(1933): 337, 373.

Six

"And Then This Bum Came Along": Leaving Germany

Although these were exciting times in biochemical and physiological sciences, by late 1932 Gerhard had some unsettling personal experiences. One occurred after he had stepped out of the hospital for a while to watch a political parade sponsored by the "Iron Front" (Eisener Front), a coalition of anti-Nazi, anti-monarchist, and anti-communist groups led particularly by Social Democrats, which had been formed in December 1931.[1] The Iron Front organized a parade in Frankfurt. It passed just three blocks from the hospital. Curious to see how much popular support the democratic parties still had, Gerhard stepped out for a while, watched the parade and then went back to the hospital. A day or two later, he was called to the office of a faculty member in the Pathology Department, a Nazi Party member, who told him, "Schmidt, you have to be very careful these days." In response to Gerhard's question, "Why?" he said, "You have been seen yesterday standing on a speaker's platform in front of the Opera House and addressing the Association of the Iron Front." Gerhard responded, "It's not true; I saw the parade. I was interested to see how many followers they had, but I never participated any way actively in it." The faculty colleague told him that a young postgraduate fellow in the department, who was also a member of the Nazi organization, reported having seen him there.

Talking of the Iron Front brought to mind another episode, which indicated that politics had entered the consciousness of scientists. Late in 1932, at the previously mentioned grand meeting of the Kaiser Wilhelm Gesellschaft in Heidelberg, the chairman for one of the sessions was Carl Neuberg. He introduced the final paper, to be presented by Otto Warburg, the champion of iron-dependent oxidative biochemical reactions in cell respiration, with the comment: "Now let's see what the Iron Front has to say."

Scientists were indeed very aware of the political scene, but most of them hoped the danger would pass. In 1932, Embden declined an invitation from the professor of medicine in Basel to become successor of Karl Spiro, the professor of biochemistry in Basel, who had died that year. Embden told

---

[1] R. J. Evans. *The Coming of the Third Reich* (New York: Penguin Books, 2004), 290.

Gerhard that he had thought about the invitation for a long time because the political situation in Germany was quite uncertain. But he had come to the conclusion—this was 1932, before he had discovered phosphoglyceric acid—that he wanted to stay in Germany to his end. He knew that some other colleagues considered the situation to be very serious. For example, Rudolf Höber, a renowned physiologist and Professor at the University in Kiel, though not Jewish, earned the enmity of the Nazis in 1931, when, as Rector of the University, he served on a court that disciplined Nazi students who had disrupted a church service being conducted by a Lutheran clerical scholar and university chaplain. When the Nazis gained power in 1933 they restricted Höber's access to his lab and then forced his retirement; in September 1933 he left Germany for England and, eventually, for America.[2]

As soon as Hitler was appointed Reich Chancellor on January 30, 1933, Gerhard knew there was no future for himself in Germany. He was not alone in this understanding; both he and an un-named colleague, an associate professor of physiology who happened to be ill and whom Gerhard visited in the hospital, came to that conclusion as they talked that same evening.

A faculty assembly, a devastating experience, was called shortly after Hitler came into power and had begun installation of Nazi Party members in all administrative posts, removing any members of other parties. This was part of the process of Gleichschaultung: bringing things into line, a forced "coordi-nation."[3] Local leaders, Gauleiters, all Nazis, were appointed at many levels. One of them, Super-Gauleiter Jacob Sprenger of the Frankfurt district, called this academic assembly together. Every faculty member received an invitation saying that the new Gauleiter wanted to introduce himself to the university, to the faculty of Frankfurt University, and explain his program. So every faculty member was invited, with some urgency, meaning you had better come; it will be noticed if you're not there. The whole faculty was invited to the Römer, the old City Hall in Frankfurt, a beautiful historic building where, in the Middle Ages in Germany, the German emperors were crowned.

In Frankfurt the faculty was very proud of its contributions to medicine. After all, in its history, even before the university had been founded, there had been Paul Ehrlich, the discoverer of arsphenamine and so much more;[4]

---

[2] W. R. Amberson, " Rudolph Höber: His Life and Scientific Work," *Science* 120(1954): 199–201.

[3] Evans, *Coming of the Third Reich*, 375–90.

[4] Paul Ehrlich (1854–1915) became Director of the Royal Institute of Experimental Therapy, when it was established in Frankfurt in 1899. He proposed far-sighted models of how immune responses occur and pioneered the concept of developing drugs for specific targets. One such drug, arsphenamine, also known as Salvarsan, was the first effective pharmaceutical treatment for syphilis. He shared the 1908 Nobel Prize for Medicine or Physiology with Ilya Metchnikoff. From *Nobel Lectures, Physiology or Medicine 1901–1921* (Amsterdam: Elsevier, 1967). Retrieved online October 23, 2012, from "Paul Ehrlich — Biography" Nobelprize.org. http://www.nobelprize.org/nobel_prizes/medicine/laureates/1908/ehrlich-bio.html.

there had been the very famous pathologist Karl Weigert,[5] and many more of great stature. In his talk, this Gauleiter said to this proud faculty that now the main thing is to make the science alive, with a new spirit, a truly German science. One thing that must be eradicated, root and stem: the Jewish parasite, which destroyed a German civilization from the inside. This was the style of his whole talk.

"In fact, it was not just shocking what this bum had to say, but you have to imagine the atmosphere. Here was this very solemn hall—in the Römer. The whole faculty of the university: the rector, the administrative head of the university—the president, who was one of the faculty members—the whole faculty. The medical faculty had many Jewish members. The pathologists were Jews—not only I—people to whom you, when you talked to them, professors, you were very cautious not to say anything that you could not really support with some facts. They had to listen to these vulgarities. They couldn't contradict, couldn't reply. They had to sit still. This ugly, really, *rain* of garbage would fall onto them. I mean—can you imagine you would be invited to something knowing that there would be nothing else but insults heaped at you, with the silent understanding that you better keep quiet or you end in a concentration camp or something like that. This was a moment of historic humiliation for the faculty. Leaving the hall, nobody dared even to talk most polite words, even implying some—even some slight disagreement with what was said. You can't imagine the shocking effect of this. The impertinence of the head of the new government! It is very difficult. Consider this together with the usual—one can call it— 'academic conceitedness' of the German intellectual. I mean the pride on being an intellectual was a marvelous thing in Germany."

Through Gauleiter Sprenger, the regime brought other changes as well. "On the benches of the park—imagine it at the Boston Common—there were labels, since the Gauleiter Sprenger came: posters [which read] 'Jews not permitted to sit here.'" When one wanted to buy a ticket for a play in a theater or for a concert, you were asked at the Cashier's Office whether you were Aryan or not. And if you said you were Jewish, then "We are very sorry, we cannot sell you a ticket."

Following the burning of the Reichstag building in February, an act of arson by a young Dutch anti-Nazi fanatic, Hitler called new elections for March 3.[6] Though he did not win an outright majority, the opposition was divided and totally ineffective and he remained Chancellor. The elaborate

---

[5] Karl Weigert (1845–904) was appointed professor of pathological anatomy at the Senkenbergsche Stiftung in Frankfurt in 1894 and received the title of "Geheimer Medizinal-Rat" in 1899. He wrote on many subjects, including staining of bacteria for microscopy; I. Singer and C. Adler, eds.,*The Jewish Encyclopedia*, Vol. 12 (New York: Funk and Wagnalls, 1906), 482.

[6] Evans, *Coming of the Third Reich*, 328–49.

inauguration of his chancellorship on March 5 "was different from any other. There were many Chancellors before—you know: Brüning, Schleicher, Papen; they became Chancellors with not much ado. We were accustomed to that. But when Hitler became Chancellor, there was an inaugural and this scene was, of course, broadcast over the radio. The whole faculty, the whole university, was invited to the main convocation hall of the university to hear this inaugural ceremony."

The medical faculty had its own convocation hall because it was separate from the rest of the university; it was in the hospital area. The whole faculty was in academic cap and gown. And there were the usual speeches. "As you can imagine, they were rather general and vague. The Führer himself said a few words. Since I had become Instructor in 1931, I was with the faculty on the stage and saw, in the audience, very disturbed faces of friends of mine. At the end, there was the national anthem, which had to be attended standing, and you can also imagine the devastating impression this whole celebration made, which meant the abolition of freedom of thought, freedom of expression. This academic affair, by the way, was the reason why I almost took a vow never to appear in cap and gown at any academic assembly, since I had the experience that cap and gown do not protect you from the most vicious official slander possible. It was really a travesty of an academic assembly."

There were some other disturbing events before this. Gerhard recalled the deprecation of Albert Einstein by the physicist and Nobel Prize recipient Philipp Lenard, the discoverer of cathode rays. Lenard wrote a set of volumes on "German Physics," in which theoretical work of Einstein was opposed and dismissed as "Jewish physics."[7] Gerhard said "Lenard was a rabid anti-Semite. He organized the dirtiest campaign against Einstein, while he was professor at the University of Heidelberg for physics. And he was himself an uncontested cheat. These happenings were known, but they were brushed aside as crazy. But this was very real. This was the climate in Germany for the near future."

Soon, Gerhard himself experienced more alarming personal examples of repression. On one occasion, he had to borrow some reagent from the Department of pharmacology. He was a close personal friend of the professor of pharmacology, Dr. Werner Lipschitz, with whom he played together in a musical quartet. Lipschitz invited Gerhard into his office and told him how shocked he was to read that morning that Bruno Walter had just been dismissed

---

[7]G. Weissmann, "X-ray Politics: Lenard *vs.* Röntgen and Einstein," *FASEB Journal* 24 (2010): 1631; available at http://www.fasebj.org/.

as director of the Leipzig Philharmonic Orchestra[8] and he made some comments about the barbarity of the government.

A few days later, Gerhard was confronted by the same Department of Pathology colleague who had spoken to him after the Iron Front parade, who now warned him that German professors will learn that they cannot openly criticize the present government; they are not in the emperor's field now. "I played dumb and said to him, 'Who would be such a fool to criticize the Hitler government now that Hitler is in power anyway?' 'Oh, but I'm not fooled in our faculty. I know exactly last Tuesday that the Professor of Pharmacology quite openly complained, in this department, about the dismissal of Bruno Walter by the Nazi government.' I knew now that our conversations had been somehow reported to this member of the pathology department. So I told him, 'I think this is absolutely impossible. More stupid things are reported about people in departments right now, politically they announce stories that are invented.' I remember he said, 'No, this story was not invented. It was reported. I heard it yesterday evening. I was at the meeting of the National Socialist section of the instructors and professors of the University of Frankfurt and the Associate Professor of Pharmacology, who is a member of our group, reported this story about Professor Lipschitz'. . . . So this conversation was reported by the Associate Professor, by the only Associate Professor of the department, next highest position to the chairman, about his own chief, who didn't know that his Associate Professor was a member of the National Socialist Party. . . . It shows that you couldn't trust anybody. . . . This was absolutely a new thing in Germany, because under the Weimar Republic you could express your opinion very frankly, even if you were an opponent to the current government. It was unheard of that any political opinion—any criticism of the government —was dangerous for your own position. . . . The fact that this person—second highest position in the department denounced his own chief officially, to have a record as an anti-Nazi, was one of the most shocking experiences I had because it was completely unheard of."

DISMISSAL, FAMILY TRAGEDY, AND DEPARTURE

By early March, Gerhard approached the chairman of his department, Professor Fischer-Wasels, and asked whether it might be good for the department if he, Gerhard, were to leave. Fisher-Wasels, responded "Leave that to me, please. In this department, I am the man who says who is leaving and who

---

[8] This comment probably refers to when Bruno Walter returned to Germany to take up his role as conductor of the Leipzig Philharmonic but was prohibited from leading a concert there on March 16, and a few days later, was prohibited from leading a performance in Berlin, *New York Times*, March 17, 1933.

is going and I want you to stay." "Well, ten days later his secretary called me. My chief wanted to have a talk with me. Then he said, 'My dear Schmidt, I know I told you ten days ago, two weeks ago, that you should stay on; but I'm terribly sorry, it is not possible any more, because I just have instructions from the government. Your contract with the city of Frankfurt [which] has to be renewed every two years ends the first of April—or the end of March—and the city refuses to renew this contract.' So, I was dismissed, with two weeks notice."

It was clear to Gerhard what he had to do in two weeks. He had to leave Germany. Fortunately, he had a destination that fit with his scientific interests. It was his good luck that, after his nonproductive visit to Helgoland in the summer of the previous year, he had applied for and received space at the Experimental Station in Naples, and by chance he had applied specifically to begin April 1, 1933, precisely when he needed a place to go, even if it was not certain what would follow.

He had to abandon his laboratory, and began to write letters, trying to identify a place to work after the month in Naples.[9] He ordered a train ticket for April 1. On March 29th "I was shocked to see by the headlines in the paper, that there must be separation, that Jewish children could not go to ordinary schools. They could only be instructed by Jewish teachers in Jewish classes in their community. All Jewish children had to leave their schools. I had a sister, my youngest sister, who was twenty-two years younger than I—so she was nine years old. This was in the headlines, that big! Of course, you can imagine the harsh style of these instructions. And the other thing which was in the headlines on that day was that, owing to the fact that Jews in other countries had protested the discrimination against German Jews that was then in full swing—as protest against these insinuations of other countries, if you can imagine, there was to be a complete boycott of Jewish stores, Jewish physicians, Jewish lawyers on the first of April. No Jew was permitted to go to a store, no ordinary citizen was permitted to—or they were warned not to enter Jewish stores. No physician should go to his office, no lawyer was permitted to go to his office."[10]

Gerhard saw these headlines on the morning of March 29, a day he planned to use to visit and take leave from a few friends. His first visit was to Embden's office. "He wished me the best, and then wrote a check for 150 marks in case—for any emergency—which I found very nice. Unfortunately, I don't

---

[9]The description by Gerhard of these last days in Germany differs in significant respects from that presented in the National Academy of Sciences biography: see H. Kalckar, *Gerhard Schmidt, Biographical Memoirs of the National Academy of Sciences*, Vol. 57 (Washington, DC: National Academies Press, 1987) 402–03. Online access at http://www.nasonline.org/publications/biographical-memoirs/memoir-pdfs/Schmidt_Gerhard.pdf.

[10]See also Evans, *Coming of the Third Reich*, 431–36.

have that; I never cashed it." In the afternoon Gerhard went to the house of the concert master of the Frankfurt Symphony Orchestra, with whom he had played in a quartet and who had always been very welcoming and pleasant.

Late in the afternoon, he stopped at the Embden home to say good-bye to Mrs. Embden. Gustav Embden opened the door and came out. He had received a telegram from Gerhard's home in Stuttgart—his mother or someone from his home had tried to call Gerhard by telephone, but could not reach him. "Anyway, he told me that my father had died." The circumstances of the sudden death of Julius Schmidt at the age of 61, are not clear. Gerhard's mother, Isabella, was away from the house, visiting a friend. The maid found Dr. Schmidt dead. The website of the University of Stuttgart, on the page on its history, states that his death was a suicide.[11] In a letter written in 1991, Gerhard's youngest sister, Renate, wrote that "Julius Schmidt died of a massive brain hemorrhage, not because he was upset about 'the little girl' not being able to go to German schools any more (I attended the Königin Charlotten Gymnasium in Stuttgart until November 12, 1938), but because March 29, 1933, was the day of the first Jewish boycott by the Nazis. All Jewish professors and teachers were locked out of their offices and laboratories and could henceforth not teach any more, Jewish students were beaten up by Nazis, Germans were not allowed to buy at Jewish stores, etc. etc. I clearly remember coming home from my piano lesson on March 29, 1933, and seeing my father at home in the early afternoon, a very unusual thing; then, later in the afternoon much commotion in our apartment, then being told in the evening that he was dead." Isabella later told Gerhard that Julius, who was not a highly emotional man, was terribly shocked by the facts stated in the headlines of March 29.

That evening Gerhard and another sister, Marion, then about 24 years old, a technician in the medical department of the hospital in Frankfurt, made the two-and-a-half-hour train ride back to Stuttgart. He could only stay briefly, however, because there was still much to do in Frankfurt, to which he returned the next day so he could settle various things in the lab.

On his way back to Frankfurt, he committed what he said was "almost high treason." He stopped at Heidelberg and went to the Kaiser Wilhelm Institute and asked whether he could see Professor Meyerhof. Gerhard told Meyerhof that he had been dismissed—not for professional reasons, but because he was a Jew. Meyerhof was also a Jew, but the Kaiser Wilhelm Institute was a private foundation, a different category than the universities, and he had not been dismissed. Gerhard told him that he planned to go to Naples, but wondered whether there might be a chance of having a job at

---

[11] Retrieved October 23, 2012, from http://www.uni-stuttgart.de/ueberblick/historie/index.en.html.

the Institute after that. "So Meyerhof was very nice and said, 'Yeh' —first he said, which was also very characteristic—I think he didn't quite grasp that I was with Fischer-Wasels. He associated me with Embden, of course. He said, 'Well, I think really Embden didn't take a firm enough stand. There are examples that, if you really show some backbone, that you can achieve something with—even with our present government. For example, I know,' he said, 'that Sauerbruch,' the most famous surgeon in Germany, in Berlin, Professor of Surgery in Berlin, 'was able to—I don't know—to keep this or that assistant in his clinic in Berlin.[12] And, perhaps, it might not have been necessary to dismiss you.' And he said, 'Well, I will— you, of course, would have to work with some of my people— might be Karl Lohmann—with Herr Lohmann—but it might be we can really do something with your enzymes. Please write to me. I think something might be arranged.' So he was—I never dared, of course, to tell Embden that I had looked for a job with Meyerhof. Even under the circumstances, he never would forgive me for that." Nothing further happened on that front.

Back in Frankfurt that day, he went to the laboratory. Fischer-Wasels was most sympathetic. Gerhard told Fischer that he had to go back to Stuttgart for his father's funeral the next day. Fischer replied that he had just called and talked with Professor Embden, whom he had invited to drive to Stuttgart with him to attend the funeral, and he offered to take Gerhard along with them. As chief pathologist, Fischer had a car and driver. Fischer also said: "And for this evening, it is not good for you that you are alone. We have some people from the faculty in our house and, if you don't object because this is—of course, your father died and you might not go to a party, but these are unusual circumstances, please come with us to our home." Gerhard accepted. At Fischer-Wasels's house that evening were several faculty department heads. The next day Fischer-Wasels, Embden, and Gerhard went together to the funeral of Julius Schmidt. They drove to Stuttgart, which is not quite as far as from Boston to New Haven. "You have to consider at that time, it was already not quite without a certain risk to ask a Jew to your home."

After the funeral, Embden and Fischer-Wasels drove back to Frankfurt. Gerhard also had to make one more trip back to Frankfurt, which was the origin point for the train ticket he had purchased earlier for the trip to Naples. Furthermore, his baggage, light as it was, remained in Frankfurt, so after the funeral he took the train back to Frankfurt. He also hoped to retrieve some DNA he had prepared from thymus gland; he wanted, at least, to precipitate

---

[12] Ernst Ferdinand Sauerbruch, not Jewish, was head of the Department of Surgery at Berlin's Charité Hospital. He opposed the Weimar Republic and supported the National Socialists, though he showed no personal animosity to Jewish colleagues and tried to stop some of the Nazi euthanasia programs; M Dewey, U. Schagen, W. U. Eckart, and E. Eva Schönenberger, "Ernst Ferdinand Sauerbruch and His Ambiguous Role in the Period of National Socialism," *Annals of Surgery* 244(2006): 315–21.

it and put it in ethanol. So he went from the funeral to the laboratory, still wearing a cutaway and top hat—something he may have owned from when he gave his public lecture at the time he became a member of the faculty. He would also use this opportunity to finally say good-bye to his chief, Fischer-Wasels, and to thank him for his incredible thoughtfulness.

This was the first of April, the day of the anti-Jewish Boycott. "I went to his office, and he was shocked. He said, 'You, here! Don't you know that just a few minutes ago the Gestapo left our department in order to check whether there are—whether any Jews had come! They would have arrested you immediately, and I would have been in great trouble. Gehen sie. Go as fast as you can, and don't leave through the main entrance of the hospital.' So I left."

The houses of some professors of medicine were in the hospital territory, and Gerhard was on very friendly terms with the son of the professor of obstetrics, so he could go though these houses, which he could enter in hospital territory but exit directly onto the street, away from the main entrance of the hospital where there were very likely some Brownshirts to check whether any Jews came out. In his nervousness and excitement, he left his top hat behind.

From the hospital he went to his rented room, retrieved his suitcase, cello, and knapsack—and boarded the train. He kept just what he could carry, including about 250 marks (about $60) from his bank account. He left behind books and any other possessions, lacking both money to ship them and knowledge of where he would be going from Naples.

The journey went smoothly. Fortunately he still had a valid German passport obtained a year earlier for travel to the Physiology Congress in Rome. He had feared that the Nazis would take away passports from the Jews; but at that time, customs officials at the border were not yet all Nazis. Nothing was examined beyond the routine superficial customs check at Basel. Nobody asked him how much money he took; so if he had had money, he could have taken whatever amount he wanted.

During the passport check at Basel, Gerhard saw a woman, a physician who had been a resident at the hospital in Frankfurt, whom he knew very well, and who had taken the biochemistry course when he was an assistant in Embden's laboratory. She too was leaving Germany at this relatively early stage. The only other Frankfurt refugee he recognized, she left the train in Switzerland, and eventually left Switzerland to go to Chicago, where she became a practicing physician.

Crossing the border, with a pessimistic outlook for Germany's future and great uncertainty about his own, he felt he was leaving his native country for good. It was a depressing time. He had just achieved the coveted goal of becoming a faculty member in a major university medical center, and knew he would not easily find a comparable position abroad. Aware of how difficult it was for a foreigner to obtain a faculty position in Germany, he projected that situation to all other countries. Somewhat later he was surprised to learn that the United States, then under the Franklin Roosevelt administration, was more open to accepting skilled people from diverse places.

Gerhard stopped overnight in a little Swiss village, Adelboden, in the Bernese Alps, to write some letters. One was directed to P. A. Levene at the Rockefeller Institute in New York, because Embden had some correspondence with him and Levene had sent some authentic yeast adenylic acid for Gerhard's research. Levene was aware of Gerhard's published work, and in fact had cited it in an American Chemical Society monograph, the most authoritative review on nucleic acids of the time, which he published in 1931. In that review, he recognized Gerhard as a student of Embden who had found a specific *deaminase*.

He wrote one more letter in Adelboden. It was directed to Professor Hans von Euler in Stockholm, to whom Gerhard had been introduced when von Euler once visited Embden's lab.[13] Gerhard had not had time to discuss potential placements with Embden in the short period between his dismissal and his departure, and he did not want to bother Embden, who, having a Jewish father, may himself have been in danger. But Embden had given him a general letter of recommendation in 1929, when Gerhard left the Embden lab to go to Fischer-Wasel's department. This would serve as a letter of introduction to a potential lab chief, who could then contact Embden for more detailed information.

TEMPORARY RESPITE IN NAPLES

The night at Adelboden was his only stop as he passed through Switzerland, into Italy, and on to Naples. In Italy, although he had felt some disgust in response to the cult of Mussolini's rule, he did not fear the kind of fascism he had witnessed there in 1932. There were black-shirted fascists, but most people did not seem obsessed with politics. In 1932, racial hatred was not a part of the government's program in Italy as it was in Germany, and Jews were not under particular pressure there. Besides, "For any person of the educated German middle-class who went through the education necessary for the university, a trip to Italy was a dream. This was still very much reinforced by the fact that Italy had influenced Goethe tremendously; Goethe made four

---

[13] Hans von Euler-Chelpin was Professor of Organic Chemistry at the Royal University of Stockholm and Director of Vitamin and Biochemical Institute in Stockholm. He shared the 1929 Nobel Prize in chemistry with Arthur Harden for work on alcoholic fermentation. See Chapter Nine.

trips to Italy. Some of his greatest works were tremendously influenced by Italy, the classical history, the whole tradition of humanism. . . . The prospect of living for a while in Naples, not even a pleasure trip, but just for your work—and being 'forced' to go to one of the most beautiful places—was so attractive that for the moment, this repressed the worries which were there about my further future." He encountered interesting people and beautiful vistas. He met at surgeon, a Dr. Capaldi, from whose home—a marvelous house in Via Tass—one could see the whole Bay of Naples, Capri, and Vesuvius, which was still smoking at that time. He met with fellow scientists over lunch, in a common dining room, which was decorated with frescoes of a very well-known painter, Hans von Marées, a German-born artist who had moved to Italy.[14]

Although he had space and a lab at the research station in Naples, Gerhard had no fellowship or other financial support, and the space assignment was for only one month. At that time, his 200 marks would be enough to live on for a month. Fortunately, his Frankfurt chief, Fischer-Wasels, sent a check for 180 marks to Gerhard's mother, who forwarded it to Gerhard. Fischer-Wasels continued to send a check for each of the next three months. Living was very inexpensive, and Gerhard could live very happily with very little money. One could buy cigarettes by the piece, "so if you didn't have much money, you always had one centisimo for a cigarette."

It was easy, without having a reservation, to find an inexpensive pensione where he could stay for the first week in Naples. He did not speak any Italian then, but learned it quickly because he found it so similar to Latin. When he arrived at the marine biology station, he found some well-known German scientists there, including a geneticist, Richard Goldschmidt, from Berlin.[15] He soon found good friends in a botanist from the institute of Frits Warmolt Went in Holland, where auxins had just been discovered, and an American biologist from Pasadena, Albert Tyler, who remained at the Naples lab for many years. Tyler, an admirer of the German humorist Wilhelm Busch, and his wife were particularly welcoming to Gerhard. Some of the German refugees introduced him to an Italian pediatrician, Doctora Mazzeo, and his wife, a very smart woman from Berlin; he retained a long friendship with this family.

---

[14] Hans von Marées, born in Elberfeld, Prussia, 1837, studied in Berlin, and went to Italy in 1864. He was most known for the frescoes painted at the Stazione Zoologica Anton Dohrn at Naples (Marine Biology Lab). These paintings, and others by this artist, are described in L. Ettlinger, "D. Hans von Marées and the Academic Tradition," *Yale University Art Gallery Bulletin* 33, no. 3(1972): 67–84.

[15] Goldschmidt was forced to resign his position at the Kaiser-Wilhelm Institute in Berlin. In Germany he had studied the mechanism of sex determination in embryological development. He moved, via Italy, to the United States, joining the faculty at the University of California, Berkeley. His idea that evolution can occur with discontinuities caused by large-scale mutational events as well as by gradual accumulation of small mutations created some controversies, but he became a member of the National Academy of Sciences and President of the Ninth International Congress of Genetics; M. Dietrich, "Richard Goldschmidt: Hopeful Monsters and Other 'Heresies'," *Nature Reviews Genetics* 4(2003): 68–74.

A few days after Gerhard arrived, Goldschmidt told his colleagues of news that Otto Meyerhof was in great danger of losing his position, in spite of a trip to Berlin made by Franz Knoop to appeal on Meyerhof's behalf. Then, two weeks after his arrival in Naples, Gerhard received Levene's answer to the letter written during the journey to Naples: that he was sorry he couldn't do anything for him because the Rockefeller Institute and the Rockefeller Foundation recently had very bad luck with their investments. They had invested heavily in railway stocks. On the positive side, Gerhard also received, in Naples, an offer from Hans von Euler inviting him to come to Stockholm as soon as he wanted. He accepted the offer, and was able to enter into his research in Naples knowing he had a next step available.

With a large laboratory, an aquarium, a pet octopus, and the indispensible accessory of every lab—a folding reclining beach chair—at his disposal, he began to collect *Arbacia*, the common sea urchin of Naples, and the less common *Sphaerechinus granularis*, the purple sea urchin. He found conditions for complete extraction of purines and purine-containing compounds. He estimated purine nitrogen to comprise about 3.5% of total nitrogen. He also found, to his surprise, that virtually all the purines—"it was at that time by no means a banal finding"—were present in the acid-insoluble fraction, which he interpreted to mean as polymerized nucleic acids, rather than low molecular weight compounds, such as nucleosides or nucleotides. Further, he noted a strong Feulgen reaction, characteristic for the deoxyribose of DNA, in the acid-insoluble material. This finding was opposed to a result that had just been published by Jean Brachet, later a renowned biochemist and cell biologist, who reported in his 1933 paper the presence of only RNA—little or no Feulgen reaction of DNA—in the unfertilized egg and postulated that this reservoir of RNA was a precursor for the synthesis of DNA during development.[16] Gerhard felt that Brachet had erred in his estimation, using a pentose-specific reagent, of a large drop in the amount of RNA during development. These measurements of very little free soluble nucleotide and the presence of DNA in unfertilized sea urchin eggs were the main results of Gerhard's work at the Experimental Station in Naples. He submitted a manuscript later in the year and it was published in 1934 in *Zeitschrift für physiologische Chemie*.[17]

Gerhard had applied for rental space at the Stazione Experimentale for just four weeks. He continued to work there beyond that time, but eventually, after a second month, Dr. Dohrn, the Director of the Aquarium, reminded him that his lease was overextended, and he had to get permission from the

---

[16] J. Brachet, "Recherches sur le synthèse de l'acide thymonucléique pendant le développement de l'oeuf d'oursin,"*Archives de Biologie, Paris* 44(1933): 519–76.

[17] G. Schmidt, "Über die Bindung der Purinbasen im unbefruchteten Seeigelei," *Zeitschrift für physiologische Chemie* 223(1934): 81–85.

German government to continue to work there. At that time, such permission was very unlikely to be forthcoming. Fortunately, some people from the University of Naples who worked at the Aquarium introduced Gerhard to the head of the Biochemistry Department of the University, Enrico Quagliariello, who immediately invited him to continue his work in that department. Quagliariello provided a room and offered to requisition any needed materials from the Aquarium on his behalf. This move was necessary because, although the Aquarium was originally a private foundation and property of the Dohrn family, the government took it over during the First World War and appointed Dr. Dohrn as lifetime Director. In the developing political environment, with the possibility that Mussolini and Hitler sooner or later would become friendly, everything had to be avoided that might embarrass Dohrn, who was not a native of Italy. Dohrn could not have risked permitting any foreigner to work there without the approval of his home country's government.

More sad personal news also came to Gerhard. A few weeks after he arrived in Naples, his mother informed him, by letter, that his best friend, a medical resident at Göttingen, had died of typhoid fever, which he had contracted during an epidemic.

Better tidings came shortly after that. He received a letter from his mother, so recently and unexpectedly widowed at about 55 years of age, saying that she wanted to visit him. Italy was the only country for which she still could get an exit visa from Germany, and as she knew that he would not stay for long in Italy, she wanted to see him, and she would use the occasion to see the country as well. She made a hotel reservation in Rome. Gerhard made the trip to Rome, about three hours by train, where she arrived, after a seven- or eight-hour train ride from Stuttgart, around ten o'clock at night. From the train station they went to her hotel. She put her luggage on the bed and, with no stop for rest or food, she said, "I have to see the Colosseum and the Forum." So they walked to the Forum and then sat in the Colosseum, tremendously impressed in these great ruins. The visit was a 'marvelous experience' for Gerhard. While they were together, his mother explained how she was able to send him a check each month. It had come, as noted above, from his former chief, Fischer-Wasels, who used research funds for this support, but he directed it through Gerhard's mother, because a family was, at that time, still permitted to send a certain amount of money privately to a relative abroad. She had not been able to tell about the real source when she sent the check, because she assumed the mail would be monitored. The 180-mark check came just with the comment: "Söhnchen, please accept it. I will tell you later, when we see each other, how I could send you this money." These checks, received over three months, provided amply for Gerhard's needs in Italy.

Another piece of more disturbing news arrived during the summer. In August, Gerhard's mother wrote to tell him that Gustav Embden had died. This news was shocking, as Embden had seemed to be in good health when Gerhard took his leave and said goodbye at the end of March. From his mother's or other letters from Germany, he learned that Embden had been at a convalescent rest home directed by a former scientific collaborator, Dr. F. Kalberlah. The pathologist Fischer-Wasels wrote to Gerhard also, saying the cause of death was a blood clot.

Only much later, at the International Congress of Biochemistry in New York in 1964, did Gerhard learn more, from someone close to the events after Gerhard left. He received a phone call, at his hotel room, from Hans Jost, one of three key members of Embden's lab in 1933. Jost had been an Associate Professor in Embden's Department of Biochemistry. His affiliation with the Nazi Party allowed him to keep his position and eventually to become Chairman of Biochemistry at Innsbruck after the Anschluss of 1938; but it also led to his dismissal from his professorship and loss of pension eligibility after the war. In spite of his political affiliation, Jost had always been friendly toward Gerhard, whom he knew to be Jewish.

Jost, in his 70s in 1964, met Gerhard at a bar one evening during the Congress in New York and told him about what occurred after April 1, 1933. That was a university vacation period between academic terms. April was also the month in which the "Law for the Restoration of Professionalism in the Civil Service (*Gesetz zur Wiederherstellung des Berufsbeamtentums*)" was passed, meaning the dismissal of Jews from posts in public universities.[18] Embden, who had a Jewish father but a non-Jewish mother, inquired of the Secretary of Education for Prussia, responsible for Frankfurt, whether he should announce his lectures for the summer term. At that time, before the Nuremberg Laws of 1935, the definitions of Jewishness had not been clarified. He was told that, for the time being, he should announce the lectures and await further instructions. Meanwhile he was invited to Basel, Switzerland, where, in mid-May, he delivered a talk on his landmark discovery and proposed pathway for glycolysis.

Embden then began his summer-term lectures. As was the practice, he was accompanied by the whole staff of the Department, including Jost, Lenhartz, and Deuticke. Jost recalled that, although none of them was Jewish, they always made a point of being with Embden in the lecture in case any unrest developed. As they entered for the first lecture, they saw all the students were in Brownshirt uniforms with swastikas. The first lectures went without incident; but then, in the middle of a lecture two students near the front

---

[18] The law was passed April 7, 1933; Evans, *The Coming of the Third Reich*, 437–40.

started a private conversation, in a whispered tone. Embden could not tolerate that and, in his most formal, quiet, humble manner, according to Jost, asked, "May I request the gentlemen in the second row to continue their conversation after this lecture?" Jost said that, after this remark "All hell broke loose in the lecture hall," with students shouting all kinds of insults at Embden, who was completely unprepared for anything like that. His staff accompanied him out of the lecture hall and to his office, where he "had a nervous breakdown." He refused to lecture any more and was in such condition that it was decided that he should spend some weeks in the sanatorium of his friend, Dr. Kalberlah. There he died on July 25.

An additional shock was delivered when Gerhard received, probably from his mother, a copy of the Embden obituary in the *Frankfurter Zeitung*. Instead of describing his landmark discoveries in metabolic chemistry culminating in solution of the glycolysis pathway puzzle, the obituary emphasized his discovery of Recresal, an inorganic phosphate "pep" pill.

Gerhard was in Naples for five months, from the first of April to early September. He worked on sea urchin eggs throughout this period and on writing some papers on his last work in Germany. On May 15 he submitted the paper on the microestimation of purines in tissues to the *Zeitschrift für physiologische Chemie*, early enough so that it was published in 1933.[19]

He completed a second manuscript only after he left Naples; it was submitted from Stockholm, December 20, also to the *Zeitschrift für physiologische Chemie*, and published in 1934. This article was his effort to have a journal paper published on the work he presented in Rome in 1932, and he had a reason to want that published in a scientific journal. His 1932 isolation of glutamylserine-phosphate was a new finding in 1932, but presentation at a Congress does not constitute publication. Other labs were also interested, at that same time, in phosphoproteins and isolation of corresponding phosphoamino acid compounds. Fritz Lipmann, then working under a fellowship in the lab of P. A. Levene at the Rockefeller Institute in New York, reported in 1932 that he isolated serine phosphate from breakdown products of the protein vitellinic acid.[20] This product was not quite the same as the dipeptide phosphate reported by Gerhard. In the summer of 1933, however, Levene and Douglas Hill published a paper on isolation of the very same glutamylserine-

---

[19] G. Schmidt, "Mikrobestimmungen von Purinsubstanzen in Geweben. 2. Mitteilung. Die Bestimmung des Guanins, des Adenins und der Oxypurine," *Zeitschrift für physiologische Chemie* 219 (1933): 191–206.

[20] P. A. Levene and F. Lipmann, "Serinephosphoric Acid Obtained on Hydrolysis of Vitellinic Acid," *Journal of Biological Chemistry* 98 (1932): 109.

phosphate that Gerhard had found.[21] Levene and Hill acknowledged that the abstract of Gerhard's Rome presentation "on the isolation of a substance identical with our own" did come to their notice when their paper was ready for publication. That statement helps to establish Gerhard's priority in the literature, although Levene and Hill also said that the Rome abstract did not give details of preparation or analytical data.

Gerhard's response was to write an article as a comment on the Levene–Hill paper.[22] In it he gave details of his preparative method, in which he subjected a kilogram of casein, in a liter of buffered solution, to a 4-week digestion with pancreatic enzyme, largely *trypsin*. His method used only enzymatic degradation of the protein, whereas Levene and Hill used both enzymatic and hydrochloric acid hydrolysis, and proceeded through an intermediate crystallization of the brucine salt of the phosphodipeptide. Gerhard commented that his totally enzymatic approach emphasized that the compound could be a truly biological product of protein breakdown. He also gave his elemental analysis, but acknowledged that Levene and Hill had a more complete characterization. He remarked that external circumstances (i.e., events in Germany) had made it impossible for him to complete the characterization that he hoped to complete after the preliminary presentation.

---

[21] P. A. Levene and D. W. Hill, "On a Dipeptide Phosphoric Acid from Casein," *Journal of Biological Chemistry* 101 (1933): 711.

[22] G. Schmidt, "Zur Gewinnung der Dipeptidphosphorsäure aus Casein. Bemerkung zu einer Arbeit von P. A. Levene und D. W. Hill," *Zeitschrift für physiologische Chemie* 223 (1934): 86. Gerhard purified the phosphodipeptide from the acid-soluble products, by repeated precipitation steps—with lead acetate first, decomposition of that product with hydrogen sulfide, and finally by formation of the barium salt, which was repeatedly precipitated with ethanol and redissolved in water.

Seven

Stockholm with von Euler

In September 1933, Gerhard took leave of Italy and moved on to Stockholm to take up his fellowship with Hans von Euler. He traveled by train through Italy, France, and the Netherlands, to the port of Rotterdam, where he boarded a freight steamer for passage to Stockholm. On the way through France, during a brief stop at Metz, he met a young woman with whom he had been friendly during his last half year in Frankfurt. They had been attracted to each other, but she was not Jewish and there was extremely grave risk for both participants in such a relationship, especially after Hitler gained power. By the time they met in Metz, Gerhard had determined that he would not again step foot in Germany, and that he had no means to support her if she were interested in emigrating. They did not meet again.

This unhappy separation was one more fractured relationship, following his father's death, his leaving his mother and three sisters in Germany, and the deaths of Embden and another close friend. He established for himself a goal: to reserve any money he could save for the priority of helping to get his mother and sisters out of Germany.

When interviewed in 1972, Gerhard reflected on the extent to which the Nazi regime affected personal lives and created an overall mood of despair. He referred to a description given by Hans Krebs, who had carried out groundbreaking research in 1931 and 1932 at the University of Freiburg, identifying urea formation as part of a cyclic rather than simply linear series of chemical reactions. Krebs, being Jewish by birth though not in practice, was dismissed from his Freiburg faculty position in June 1933.[1] His research had been noted and admired by Frederick Gowland Hopkins, a renowned professor of physiological chemistry at Cambridge University in England, who invited Krebs to join his lab and arranged for fellowship support from the Rockefeller Foundation. Thus Krebs escaped to Cambridge. He had occasion to recall those days when he presented the annual Hopkins Memorial Lecture in 1961. Near the end of the lecture, Krebs spoke of the atmosphere in Germany: "The contrast between the world from which I had come away and the

---

[1] F. L. Holmes, *Hans Krebs, the Formation of a Scientific Life 1900–1933* (New York: Oxford University Press, 1991).

world which I entered was indeed enormous. Nazi philosophy had infiltrated all strata of German life and had created a dreadful atmosphere. The teaching of *Mein Kampf* had whipped up the worst of primitive instincts. There were savage brutalities in day-to-day life in the form of beatings, shootings and murder, personal vituperation, bad faith, incredible arrogance. Before the Nazis assumed power the worst features of their creed had been somewhat restrained, in particular in the universities, but after 30 January they swept into the open everywhere. It was a great shock to see how some of one's immediate colleagues, either because of conviction or because of blackmail and political pressure, adopted the official Nazi doctrines and cold-shouldered the political opponents of the Nazis. After this experience the warmth which greeted me at Cambridge from the minute of my arrival was indeed touching and heartening. There was a spirit of friendliness, of humanity, of tolerance and of fairness which had vanished in Germany."[2]

Hans von Euler's prompt and favorable response to the letter Gerhard wrote from Adelboden en route to Naples led to a brief sojourn in Stockholm, at the Vitamin Institute and Institute of Biochemistry, which had been established in 1929 by the Knut and Alice Wallenberg Foundation and the Rockefeller Foundation, with von Euler as its director. Gerhard had the good fortune of being on his way to working with one of the world's leading biochemists; but the fellowship was for only half a year.

Hans von Euler was also of German origin, having been born in Augsburg in 1873; he retained a strong Bavarian accent after his many years in Sweden.[3] He had studied with leading scientists in Berlin: in chemistry with Emil Fischer, and in physics with Otto Warburg and Max Planck, and earned his doctoral degree from the University of Berlin in 1895. After two years of further training with the physicist Walther Nernst at Göttingen, he moved to Stockholm 1897 to work with Svante Arrhenius, considered to be one of the founders of physical chemistry, at the Royal University of Stockholm. He was appointed privatdozent at the university in 1898 and professor of general and organic chemistry in 1906.

Von Euler had carried out award-winning research on the catalysis of hydrolysis, studied the chemistry of plants and fungi, and then became one of the leaders in enzymology and fermentation, publishing major works in these fields and sharing the 1929 Nobel Prize with Arthur Harden for his work on fermentation. He clarified the structure of cozymase (now known as nicotinamide-adenine-dinucleotide; NAD), a small molecule essential for the

---

[2] H. A. Krebs, "The Physiological Role of the Ketone Bodies: The Third Hopkins Memorial Lecture," *Biochemical Journal* 80(1961): 225–33.

[3] Von Euler biographical material is from Nobel Foundation, *Nobel Lectures, Chemistry 1922–1941* (Amsterdam: Elsevier, 1966).

activity of many enzymes that catalyze metabolic oxidation–reduction reactions, work that led him to extensive studies of vitamins. In the late 1920s he showed a relationship between growth and carotene and vitamin A activity, an interest that would be a basis for some of Gerhard's work with him. By 1935, the year Gerhard left Stockholm, von Euler was studying biochemistry of tumors and nucleic acids.

The location of the Biochemistry Institute at the intersection of Odengarden and Valhallavagen gave Gerhard a "Wagnerian" feeling as he approached it. Von Euler gave him a very friendly welcome. He already had received two other German refugees. One was Eric Adler, who went on to publish major papers with von Euler on the biosynthesis of NAD and NADP.[4] The second was Bernhard Zondek, a widely known and wealthy obstetrician and gynecologist who had been dismissed from his position as chief of the corresponding ward at the municipal hospital of Berlin-Spandau and who had, with Selmar Ascheim, developed the first assay for estrogen and for pregnancy.[5] In Sweden, Zondek worked on the hormone intermedin, responsible for fish acquiring a colored line on their body at the beginning of the mating season. Later in 1934 he accepted a position in Jerusalem, where he became professor and head of obstetrics and gynecology at Hebrew University's Hadassah Hospital. Zondek, a charming and witty person, figured prominently in Gerhard's life in Stockholm. Among other things, he organized, with Adler and Gerhard, a chummy "Thursday Club" dedicated to German literature.

PURINE CHANGES IN CELL GROWTH STIMULATED BY VITAMIN A

Although Gerhard published three papers based on his research in Stockholm, he felt he did not make important gains scientifically. In his view, von Euler differed markedly from Embden in relationships with colleagues and students. Von Euler was the "Geheimrat," above his fellows, whereas Embden had been personally close to those working in his lab, considering his role as teacher ahead of his role as department chief. Gerhard would have liked to pursue independent research in Stockholm, but recalled that von Euler was

---

[4] NAD is a dinucleotide, a combination of one nucleotide made up of nicotinic acid, ribose, and phosphate covalently linked to a nucleotide of adenine, ribose, and phosphate. Nicotinamide is a nitrogen-containing base derived from the vitamin niacin. NADP has the same nucleotides as NAD but with an additional phosphate group attached to the ribose of the adenine nucleotide. Both NAD and NADP are involved in oxidation-reduction reactions (accepting or donating electrons from substrates) that are important in generation of usable chemical energy as in the synthesis of ATP or other electron-requiring synthetic reactions.

[5] V. C. Medvei, *The History of Clinical Endocrinology* (Parthenon, Carnforth, UK: 1993), 482–83; M. Finkelstein, "In memoriam. Professor Bernard Zondek," *International Journal of Fertility* 12(1967): 285–7.

interested mainly in the method for purine analysis that Gerhard had published, and expected Gerhard simply to apply it to matters of interest in the
department's ongoing research. On some days faculty members would bring
samples and ask him to determine their purine content, thereby treating him
as a technician.

Still, Gerhard achieved significant accomplishments in his short time of
just several months in Stockholm. He developed some modifications that
improved the reliability of his total purine determination method. By reducing
the amount of copper sulfate used for precipitating the purines, he reduced
the amount of contaminating protein sources of nitrogen and obtained close
duplicate values for purines whether he used small or large amounts of starting
tissue samples. The higher amounts of copper sulfate used up until then
contained significant amounts of copper oxide, which co-precipitated contaminating proteins, most notably when large tissue samples were used.

With the modified method, Gerhard measured the influence of Vitamin A
and carotene on purine content of many tissues. Because von Euler was
especially interested in the role of carotenoids and Vitamin A as growth
factors, Gerhard tested the hypothesis that the growth effect would be reflected
in the amounts of nucleic acid components needed for increasing cell numbers.
As predicted, in growing tissues, such as embryonic organs or experimental
tumors (the Jensen sarcoma), and in highly cellular organs, such as the spleen,
the purine nitrogen content was higher and made up a larger fraction of total
nitrogen than in nongrowing tissues, such as muscle or liver. Furthermore,
dietary deprivation or supplementation with carotene led to a corresponding
decrease or increase in the ratio of purine nitrogen to total nitrogen in
comparison with normal dietary intake. These results, linking carotene and
Vitamin A intake to growth through nucleotide and nucleic acid metabolism,
were published in a paper, with von Euler as first author, in *Zeitschrift für
physiologische Chemie*.[6] In this paper, Gerhard noted that there was no known
mechanism for conversion of Vitamin A to purines and one could only specu-
late on the mechanism of the vitamin's action. This comment reflects the
focus of the era on biochemical pathways as the frame of reference for
understanding mechanisms. Eventually it has become known that Vitamin A
modifies genetic regulation and gene expression, triggering activation of many
reactions required for cell growth;[7] but in the 1930s there was no infrastructure
of knowledge that could permit one to dream of such a mechanism.

Following a similar rationale concerning Vitamin A activity, Gerhard and
a coworker in the Biochemistry Institute, Inga Rydh-Ehrensvärd, measured

---

[6] H. von Euler and G. Schmidt, "Einfluss des Carotins (Vitamins A) auf den Purinegehalt wachsender
normaler und pathologischer Gewebe," *Zeitschrift für physiologische Chemie* 223(1934): 215–28.

[7] M. H. Stipanuk, ed., *Vitamin A Biochemical, Physiological, and Molecular Aspects of Human Nutrition*,
2nd ed. (St. Louis: Saunders Elsevier, 2006).

the influence of carotene on the *guanase* content of rat spleen. They reported that, consistent with a Vitamin A action that affected both growth and nucleic acid metabolism, the amount of *guanase* activity was markedly diminished in animals fed a Vitamin A-free diet.[8]

While in von Euler's lab, Gerhard also embarked on a different avenue of research, combining issues that would command his interest for the rest of his career: the properties of nucleoproteins, the varying stability of different kinds of phosphate esters, and the development of micro-methods for biochemical analysis. Von Euler had been studying mutant plants lacking chlorophyll and sought ways to understand the differences in chemical composition of normal and mutant cells. Interested in comparing their chromosomes, he needed analytical methods that could be applied to small amounts of tissues, on the order of 10 g. As a model for developing such methods, he and Gerhard chose to study the nucleoprotein of fish sperm and fish testicles, which are rich in chromosomal material. More specifically, they chose to measure the content of nitrogen, phosphorus, and phosphoprotein. For the latter, they took advantage of the finding, previously described by Plimmer, that phosphate esters in phosphoproteins are hydrolyzed on exposure to alkali,[9] yielding inorganic phosphate, whereas the phosphate esters in nucleic acid remain intact under these conditions. Gerhard developed a method[10] with which he found that phosphoprotein makes up a measurable but very small fraction of the total weight of several tissues or organs. In fish testicles, it was negligible compared to the nucleic acid phosphate. It was even lower in a metastatic liver carcinoma sample.[11]

There followed, in the same article, a section attributed just to Gerhard: a method for complete extraction of nucleoprotein and its fractionation by precipitation at varying pH. One fraction, about 10% of the total nucleoprotein, was precipitated at pH 7.2 and had a much higher phosphate-to-total-nitrogen ratio than fractions precipitated at lower pH—that is, was enriched in nucleic acid—but in all fractions he found a consistent value for the molecular ratio of purine nitrogen content to phosphate content, 5:2. This section of the article includes a statement, important in relation to his later work, that, although there is no definitive evidence on the matter, the results of this study, especially

---

[8] I. Rydh-Ehrensvärd and G. Schmidt. "Über den Einfluss des Carotins auf den Guanasegehalt der Rattenmilz," *Zeitschrift für physiologische Chemie* 227(1934): 177–80.

[9] R. H. A. Plimmer and W. M. Bayliss, "The Separation of Phosphorus from Caseinogen by the Action of Enzymes and Alkali," *Journal of Physiology (London)* 33(1906): 439.

[10] The method involved preliminary removal of inorganic and acid-soluble phosphate-containing material from the tissues with trichloroacetic; removal of lipid phosphates with alcohol and ether; dissolution and incubation of the residue in 1% sodium hydroxide for 24 hours at 37 degrees C; precipitation of protein; filtration and measurement of inorganic phosphate in the soluble filtrate.

[11] H. von Euler and G. Schmidt, "Über Nucleoproteide der Fisch-Testikel," *Zeitschrift für physiologische Chemie* 225(1934): 92–102.

the pH-sensitivity of precipitation and re-dissolution, was consistent with the idea that the nucleoprotein is a salt-like association between the nucleic acid and the protein.

He wrote one different kind of article while in Stockholm. His former friends at Embden's laboratory thought that, in spite of the political circumstances, it would be only fair that he would be one of the people to write something in memory of Embden.[12]

Considering the uncertainty of his life situation, his arrival into a new country, culture, and laboratory, and the short time available, the three scientific publications from Stockholm represent a respectable scientific output. Furthermore, patterns of his lifetime work began to emerge here. He would continue to develop new and improved analytical methods and to study phosphoproteins and, particularly, nucleoproteins. His stay in Sweden also afforded him the opportunity to meet and visit with outstanding scientists in his field of interest, including Karl Myrbäck, who had written to him and Embden with comments on their first paper on *adenosine deaminase*; in Sweden they had interesting discussions about the postulated structure of cozymase. He also visited the Karolinska Institute and met Einar Hammarsten, who had described, in 1924, preparation of the purest DNA yet available and, from the viscosity of its solution, concluded that it was a very large molecule. He also met Torbjörn Caspersson, who pioneered histochemical studies of cells and application of ultraviolet spectroscopy at the histological level, as well as staining that allowed appreciation of chromosomal substructure.[13]

ENJOYING STOCKHOLM BUT LOOKING FOR A NEXT STEP

Beyond the science, Gerhard enjoyed several other aspects of his months in Stockholm. He was not entirely without family there; a cousin, a businessman, had moved from Munich to live in Stockhlolm and worked as a manager in a department store. Furthermore, the presence of other German scientists— Adler and Zondek—and the "Thursday club" for literature provided friendship and familiar culture.

He found relatively inexpensive sustenance, taking lunches in an institution that impressed him and his colleagues: the smorgasbörd at Margareta's Husmor Skole (Home School), where girls learned to prepare the smorgasbord and

---

[12] G. Schmidt, "Gustav Embden," *Münchener Medizinische Wochenschrift* 80 (1933): 1942–44.

[13] G. Klein and E. Klein, "Torbjörn Caspersson, 15 October 1910– 7 December 1997," *Proceedings of the American Philosophical Society* 147, no. 1(2003): 73–75.

earned some money by serving guests. This smorgasbörd was frequented by business people from downtown Stockholm, and, though the quality was not the highest, the choice was enormous—everything was represented—even more than at the Hotel Kronprinz where they occasionally went on Sunday. It took him a while to follow the local procedure. As he said, "Of course, Germans have a tendency to do things systematically. We started at the left corner of that smorgasbörd and, just like the duplication of DNA, proceeded linearly without leaving out one dish; whereas the natives, the Swedes, looked over the whole plan and then made their combinations in their minds, and then took two or three tidbits of fish, and then things which fitted together and which they liked. They went there once or twice and left out most of it. I mean, out of these eighty-five dishes they took four or five, whereas the refugees . . ."

The well-off Zondek did not go to lunch to Margareta's Husmor Skole, but to a very good restaurant that was diagonally opposite the research institute—a fancy restaurant in a hotel. "Occasionally, he asked me to come over and have lunch or dinner with him. Of course, Margareta's Husmor Skole was not the place to have the best introduction into the most polished manners of Swedish society, and so I used the same technique of the smörgåsbord which I did there, without leaving the linear continuation—and I had to go back to the smörgåsbord, which was perfectly permissible, three or four times, so that usually at the end of the luncheon Zondeck, with a smile said, 'It was a great pleasure for me to have invited you again at the expense of this restaurant for lunch.'"

There were other diversions as well. He attended a dance club called the Kaos near the research institute. He found people generally very pleasant and polite, though reserved and lacking some vitality that he had greatly appreciated in Naples, even though his living circumstances were more meager in his first stay in Italy. The wife of a pathologist in Stockholm was an excellent pianist, whom he joined in playing chamber music. He had carried his cello from Frankfurt, through Naples, to Stockholm, and would continue to keep it with him in future travels, never encountering any difficulty with customs agents on the way.

It was necessary, however, for the refugees to look for the next step that would follow the short-term fellowships. Von Euler was able to invite Adler to stay on, but did not have a long-term position for Gerhard, so Gerhard wrote some letters in January 1934, including one directed to the Hebrew University in Jerusalem. Other German Jewish scientists pursued possibilities in Jerusalem as well. As noted previously, Berhnard Zondek went there in 1940. The Hebrew University had offered an appointment to Hans Krebs in 1933 and thought the offer was accepted, until Krebs was drawn instead to

Cambridge, England. Another potential destination for refugees was Istanbul. The distinguished biochemist Felix Haurowitz fled from Prague when he lost his teaching position after the Nazi takeover in 1938. He accepted the chairmanship of the Department of Biochemistry at Istanbul, where he stayed for 10 years before moving to the University of Indiana in the United States. He was not alone. The journal *Nature* reported: "Many former members of staffs of German universities and other institutes have, as is now well known, either been forced to leave or have voluntarily vacated their posts, for political, racial and other reasons. Some have obtained analogous posts in the universities of countries outside Germany, and we have recently received a list of those who are now working at the University of Istanbul. Among these are Prof. H. Winsterstein (physiology), Prof. M. Brauner (botany), Prof. M. Dember (physicist), Prof. M. von Mieses (mathematics), and nearly thirty others, most of whom have been appointed to chairs in the University of Istanbul."[14] Arnold Reisman has provided an extensive account of how German refugee professors shaped a new academic infrastructure in Turkey.[15]

Pursuing contacts that he had made while in Naples, Gerhard had also written to Francesco Pentimalli, professor of general pathology at the University of Florence. Pentimalli had worked for half a year with Otto Warburg in Berlin, and was interested in having a biochemistry laboratory in his own department. He offered Gerhard a one-year position to help set up the lab. Still waiting to hear from Jerusalem and thrilled with the chance of returning to Italy, particularly to Florence, Gerhard accepted the offer, so he did have an option for a next step after Stockholm.

He also sent an application in January 1934 to the Society for the Protection of Science and Learning (SPSL), which was trying to find positions for displaced German scholars. As described in the archives of this Society, now at the Bodleian Library of Oxford University,[16]

> The Society for the Protection of Science and Learning was founded in 1933 as the Academic Assistance Council, by a small group of academics (notably William Beveridge, Leo Szilard and Lord Rutherford). Aware of the potentially large-scale dismissal of university teachers by the Nazi régime in Germany, the council aimed to provide short-term grants for refugee lecturers, and to help them in finding new employment. This operation was funded with money raised by appeals to the academic community and others in Britain. In 1936, faced with growing demands on its services, the Council was more formally re-established as the Society for the Protection of Science and Learning, with an advisory council, executive committee, grants allocation committee, and a small secretariat.

---

[14] "German Refugees at the University of Istanbul," *Nature*, 139(1937): 747.

[15] Arnold Reisman, *Turkey's Modernization: Refugees from Nazism and Atatürk's Vision* (Washington, DC: New Academia Publishing, 2006).

[16] Retrieved Oct 29, 2012, from http://www.rsl.ox.ac.uk/dept/scwmss/wmss/online/modern/spsl/spsl.html

The Society has continued to operate, through World War II and up to the present time, having been re-named in 1997 as the Council for Assisting Refugee Academics. The SPSL file for Gerhard contains his biography (a mixture of typing and written modifications), a copy of the letter of general introduction given to Gerhard by Embden in 1929, and a laudatory letter from Hans von Euler. It also includes a letter dated May 1, 1934, from the SPSL secretary informing Gerhard that "the possibility of a position for you in Jerusalem no longer exists" and that "We have already put your name forward in connection with openings in New Zealand." A corresponding letter was sent to inform W. Adams at the Royal Society in London of Gerhard's interest in that position. The latter correspondence was signed by Ulf S. von Euler, who was then a young trainee in the lab of A. V. Hill at University College in London. Hill had been one of the founders of SPSL; Ulf von Euler, who eventually earned a Nobel Prize, was the son of Hans von Euler, Gerhard's chief in Stockholm. The letter from Embden, March 2, 1929:[17]

> Dr. Gerhard Schmidt has been working at this Institute for a long time. As a student he did scientific work here in 1922 and 1923. He spent also with us a part of his time as medical practitioner (from October 1925 till January 1926). Since that time he has been my collaborator continuously from 1st February 1926 till 31st July 1928 as a fellow of the Rockefeller Foundation, and since 1st August 1928 as a scholar of the Notgemeinschaft der Deutschen Wissenschaft.
>
> During the long years of his work on chemical–physiological problems, Dr. Schmidt developed himself more and more into an investigator of prominent qualities, and I consider him as one of my most capable and successful pupils. His scientific ability and his knowledge entitle and fit him to direct any large Institute notwithstanding his youth. His published works and original research have already made his name known in physiology. His great scientific abilities are backed by unusual human qualities. Dr. Schmidt is a man of very fine character and wide general knowledge in many fields. I regret immensely that a regular position at the Institute which I direct is not free now for Dr. Schmidt.

The letter from Hans von Euler, 18 January 1934:

> In April, Privatdozent Dr. Gerhard Schmidt accepted my invitation to participate as a scientific collaborator in the chemical investigations at the Institute of Biochemistry at the University of Stockholm.
>
> Since September 1933, he has been working on the application of his previously elaborated determination methods of purine to the investigation of spermatozoa and growing tissues.
>
> Dr. Schmidt's outstanding gifts and his keen scientific interest are evident from his previous career and work. In his work here I have found him equally zealous

---

[17]This and the following letters are cited with kind permission of CARA, the Council for Assisting Refugee Academics.

and able. Personally he is a very agreeable collaborator and his knowledge is both many sided and thorough.

I should especially like to mention his teaching ability. He is a vivid and impressive lecturer.

If it were in any way possible I should be very glad to keep him for many years here in the Institute of Biochemistry at Stockholm. Unfortunately all that I can do is to recommend him very warmly for any academic position either in medical chemistry or chemical biology.

By the time Gerhard's fellowship in Stockholm ended, the SPSL had not identified a position for him, but he was happy to return to Italy for his next steps in science.

Eight

Adventures on the Return to Beloved Italy

For his return to Italy, Gerhard first crossed Sweden by train in order to board a freight steamer in the port of Malmö, at the southwest tip of the country. The ship would travel to Amsterdam, with a 24-hour stopover in London, so he planned to see his younger sister, Marion, who had reached England in 1933. Earlier, she had been a technician in a medical clinical lab in Frankfurt. Gerhard had managed to find a position for her through a friend in Cambridge, in the laboratory of the well-known vitamin specialist, Leslie J. Harris. In Cambridge, she later met and married Dr. Ernest Childs, a scientist in the Agricultural Research Council and the School of Agriculture of the University of Cambridge. She was delighted with the possibility of seeing her brother on his voyage.

Gerhard also planned that, once he reached Amsterdam, he would travel over land, with a stop in Paris to visit the uncle of his Frankfurt friend Ernest Bueding,[1] in part to explore the possibility of a longer term scientific position in France.

The travel across Sweden began without incident. On reaching Malmö at about 7:30 a.m. one morning, with his freight steamer not due to leave until 3 p.m., Gerhard carried a suitcase, knapsack, and his cello from the railroad station to the nearby port. He passed several wharfs and came to one on which he saw a sign saying, "Next Mail Boat to Copenhagen leaving 8:15." He calculated that he had ample time to take advantage of the close proximity of Malmö to Copenhagen, just an hour and a half away by this mail boat, where he would be able to visit a family friend who might be able to help him in his search for a permanent appointment. The friend was "a very wealthy owner of a big grain mill, who was very much interested in chemistry, and had consulted very often, for the past four years, my father, who was an organic chemist. This man had his own laboratory and cultivated chemistry

---

[1] Ernest Bueding grew up and studied in Frankfurt, was displaced to Paris and Istanbul during World War II and moved to the United States, where he developed a distinguished career in pharmacology, particularly in studies of anti-helminthic drugs. Like Gerhard, he was a lover of music. H. J. Saz and R. P. Saz, "In Memoriam Ernest Bueding (1910–1986)," *J. Parasitenkunde.* 72 (1986): 697–99.

as a hobby. His name was Dr. Trönsgaard. He was particularly interested in cleavage of proteins in anhydrous media because he had found a product of this cleavage which gave the pyrrole reaction for ring compounds. He developed, on this basis, a theory, now completely abandoned, of cyclic structures being the backbone of protein structure. Since my father was a specialist in pyrroles—he had been a student of Knorr, who was particularly well-known for introducing the anti-fever drug Pyramidone [antipyrine]—Dr. Trönsgaard had been terribly nice to my parents. He invited them once on a marvelous trip, to accompany him to Switzerland. He paid all the expenses—the best hotels—for, I think, two weeks. So I thought, in the present situation, I'd at least say hello to him in Copenhagen. So I managed, in great haste, to bring my suitcase and the cello, already, to the steamboat and then run back to that wharf. I had just time to take a ticket to Copenhagen and went to the mail boat."

On the mail boat he noted a timetable and, confident in his understanding of Swedish by then, he concluded that the next return trip would leave Copenhagen at 12:50 and he would be back in Malmö a bit after 2:00 o'clock, leaving ample time to get to the boat to London. When he arrived in Copenhagen, he called Dr. Trönsgaard, who was at home and told him he was most welcome to visit. With only a few hundred kroner, about a hundred and fifty dollars, he reasoned, "Well, in my situation it doesn't pay to pinch pennies. Now I'm going to visit a millionaire interested in science, telling him that I just had been dismissed from a German university because of Hitler, and so on. There might be here an opening for me for a much better position than the offering in Florence." So he spent money for a taxi for the trip to a most magnificent house on an exclusive boulevard several miles from the port.

Trönsgaard bade him a very warm welcome, but immediately opened with, "You see we have recently moved to this modest home because business is not too good." Gerhard was "sobered in his dreaming." Continuing the conversation in German so Gerhard could understand, Trönsgaard then described his medical problems, including recent surgery for gastric ulcer, with all the details of his hospitalization and the associated pain. Gerhard never succeeded in bringing the conversation around to the political issues or his own situation. A butler entered and offered him a fine cigar (he did not smoke); and it was time to call a taxi to return to the port. He left the house with nothing accomplished.

When he reached the dock, there was no boat. After waiting ten or fifteen minutes, he asked someone about it and was told that the boat he wanted had already left—on time at 12:15, not 12:50, as Gerhard had misunderstood in his overconfident reading of the schedule. The next boat wouldn't leave till 3 p.m., meaning he would miss his steamer, with his baggage, cello, and all he owned. A possibly situation-saving idea came to him. He took a bus to the airport and asked whether there was a flight from Copenhagen to Malmö. There was, and a seat was available. It was his first flight, one of about 10 minutes. It arrived at the Malmö airport close to 3 o'clock, the scheduled departure time for his boat. The airport was not too far from the harbor, to which he rushed—whether by taxi or by foot was lost in memory. He came running up to the departure dock just as the boat was leaving, a few minutes after 3 o'clock. "The captain was looking out for me, on the landing bridge. And with the greatest politeness he said to me, "Doctor Herr Schmidt, we waited for you.""

Gerhard was the only passenger on the freighter. The trip was wonderful, somewhat more than a day from Malmö. When the steamer approached England, "The captain sent someone to tell me that, if I wanted to see the arrival of that steamer in London, I would be welcome to come to the commander bridge. I just felt thoroughly marvelous, very impressed. The captain and I were alone on the commander bridge and, particularly, I remember my greatest impression was arriving in the morning in London, a beautiful morning, on the Thames, and the steamer went far up the Thames. And at a certain moment, very impressive, a bridge came closer and closer and when the steamer was close enough to that bridge, it opened. It was the Tower Bridge. The whole traffic of London on both sides stopped at the Tower Bridge, and I felt a little Napoleonic."

On landing, Gerhard faced an immigration or passport control officer, who asked him, in a 40-minute hearing, about his origins and his aims. Gerhard showed the officer that he was indeed booked for continuing on to Amsterdam and only wanted to visit his sister in Cambridge during the ship's 24-hour layover in London. Eventually, with great politeness, the officer gave him a 24-hour pass, but with a stern warning to be back at the boat in time.

A problem arose. Gerhard forgot where, precisely, he and his sister Marion had agreed to meet. Was it in London or Cambridge? In one of his letters he had suggested that they meet at one of the corners of Westminster Abbey, but then he may have written later that he would come to Cambridge. He started by taking a bus to Westminster Abbey. He had not realized the size of that building or how difficult it would be to locate the right corner. There was much traffic and he concluded it was hopeless to try to find her there. After about half an hour, he decided to go to Cambridge; she would certainly show up at home sometime during the day. He did not understand or speak English, having had just one year of formal study in the gymnasium. Somehow he learned that the train would leave from Liverpool Station. When he got there, he did not understand the ticket purchasing system, so he asked a policeman for help. "I never saw anything that polite." The police officer took him to the ticket office, selected the right amount of money from his hand and purchased the ticket, and guided him to the train and even to a seat in the coach, while Gerhard carried his knapsack and cello.

It was getting dark in the late afternoon when he arrived in Cambridge. He knew the name of the street on which his sister lived, but not the house number. It turned out to be a very long street and one house looked exactly like any other. It seemed hopeless. Then, "I thought I would give her some signals. At that time I could whistle fairly loudly. So I walked up and down this street whistling the third cello sonata by Beethoven—the main theme—in the hope that she would hear me." The street was empty; it was eerie. Finally he saw, in the distance, a young man coming toward him, so he went toward this person to ask whether there may be rooms for rent nearby. He then recognized the young man: Hans Krebs! "Woher kommen Sie?" Krebs asked. Gerhard responded that he had just come from Stockholm. "You know, my sister is in Cambridge." "Oh yes," responded Krebs, "I'm just coming from her house. She invited several people and made a little party for you, and we are just breaking up because you didn't come."

They returned to his sister's home and had a good time together with the few who were still there. "It was one of those early spring days, and we sat around the fireplace. Our backs, of course, were cold." He had last seen his sister at their father's funeral, nearly a year before.

Early next morning, he visited Krebs in the laboratory, in the department headed by Frederick Gowland Hopkins, who had invited Krebs to Cambridge. After the narrow miss at Malmö, he wanted to be sure not to miss his boat and the officer's deadline, so he took an early train back to London, where he had some time to stroll. On the street he met a good friend from Germany, a physician from Vollhardt's clinic in Frankfurt, now a clinician at Sir Thomas Lewis's Hospital in London. He returned to the boat in time; the officer was there.

In Amsterdam, Gerhard visited an aunt and uncle who had fled from Nürmberg. After staying with them for a day, he visited a Frankfurt friend, a doctor of economics, who had left Germany for refuge in Amsterdam. Together they visited museums and enjoyed seeing works of the great Dutch painters.

Following up on arrangements he had made before leaving Stockholm, Gerhard traveled to Paris, where he stayed for eight days in the palatial home of Ernest Beuding's uncle, a businessman who lived near the Bois de Boulogne. Gerhard saw much of Paris during those days. His host also gave him a letter of introduction to André Mayer, a famous physiologist at the College de

France.[2] The visit with Professor Mayer was, however, a most disagreeable experience. Mayer seemed to still hold anti-German resentment from the First World War and had no sympathy for Germans even if they were displaced Jewish scientists. He abruptly told Gerhard that times were hard and there were no positions available, and he could not do anything for him, and sent him on his way. "I crawled out and I thought, 'Now, let's go to Italy. Let's get out of here.'" He had experienced a marvelous time in Naples and was "deadly homesick" for Italy. After this interview with Mayer, the very idea of considering France as a possibility for a later time was completely excluded.

A SECOND RESPITE IN ITALY

Gerhard's train from Paris, via Geneva and Genoa, arrived in Florence at about 5 a.m. on an April day in 1934. At this early hour, it was very quiet in the city. He walked in the vicinity of the railroad station, admired the "doors of paradise"—the Ghiberti doors of the Baptistery at the Cathedral— and, hungry and thirsty, was able to find an open-air market where he bought an orange, the taste of which remained rich in memory. By 6 a.m. he was browsing a newspaper for apartments near the university's medical school and hospital, in the countryside just beyond the district of Rifredi. That was some distance away from the main campus of the university and from the center of the city. It was reached by a major street, Viale Morgagni, named for a renowned early eighteenth-century pathologist.

By 8 or 9 a.m., he took a streetcar to the Rifredi district and located and rented an apartment on Via Marco Tabarrini. This was in a working-class suburb, a low-cost housing development of pleasant modern quarters built under Mussolini's government. The apartment owners were a couple who had two sons, who were very kind to him and very witty, sarcastic, elegant— "real Florentines."

Gerhard was delighted with both his living situation and the locale and ambience and rhythm of the medical institute, in the foothills of the Appenines. The Department of Experimental Pathology was in a building two or three stories high, with very large rooms. It had a big garden, and just outside was a marvelous shade-producing fig tree. He reached the lab by bicycle. It was too far to return home for midday break, so, while most faculty drove home for lunch and siesta, he joined those for whom lunch was prepared in the department by a cook, Artemia by name, who was also the department's cleaning woman. After delicious scallopini Milanesa and espresso, he took his siesta under the fig tree.

---

[2] André Mayer was Professor of Physiology and Vice President of the College de France. His son, Jean Mayer, became a prominent nutritionist in the United States and a president of Tufts University.

Professor Pentimalli, who chaired the Department of General Pathology of the Royal University of Florence from 1933 to 1936, was very pleasant in welcoming Gerhard to his research on Rous sarcoma virus. After spending some time with Otto Warburg in Berlin, Pentimalli was interested in biochemical studies, particularly on cytochromes, reflecting the state of respiration in transformed cells, so he asked Gerhard to establish a biochemistry laboratory. Pentimalli, apparently a high-ranking member of the Fascist Party, had no difficulty attracting government funds for his research.

Gerhard's research during his brief stay in Florence combined his growing interest in various forms of phosphorus in biological tissues and Pentimalli's interest in Rous sarcoma tumors. They took note of reports in scientific journals that the content of phosphorus in the blood of chickens with sarcomas was higher than in healthy chickens, but the various forms of phosphate had not been studied.

Gerhard was particularly interested in phosphoproteins, which he thought may be a source for the phosphorus used in nucleic acid synthesis. A method for determination of phosphoprotein had been reported in 1906 by Plimmer and Bayliss in England;[3] it was based on release of inorganic phosphate by the ready hydrolysis of the protein–phosphate ester bond by alkali. In the early 1930s few phosphoproteins had been identified—notably vitellin in egg yolk and casein in milk. Thus, Gerhard undertook measurement of total phosphorus, acid-soluble phosphorus, phosphoprotein, and lipid phosphate in the serum of eight healthy and twenty-two sarcoma-bearing hens; he also measured total nitrogen for comparison. He determined that plasma from sick hens did contain phosphoprotein, which was not detected in plasma of tumor-free animals. He found, in fact, that all forms—total phosphate, lipid-phosphate and acid-soluble phosphate—were significantly elevated in venous blood plasma from the sarcoma-bearing wing of sick hens in comparison with plasma from either healthy animals or from the nonsarcoma-bearing wing of sick animals. He reported the results in a publication, with Pentimalli as first author, in "*biochemische Zeitschrift*."[4] Time did not permit pursuit of questions about sources of the increased phosphate levels or whether the phosphoprotein was a direct product of the tumor tissue.

Thus, the several months with Pentimalli did not yield a completed scientific story or significant growth in Gerhard's own research development. On the other hand, it was one of the most pleasant periods in his life. The atmosphere at the university was warm and stimulating. He met and admired a Polish

---

[3] R. H. A. Plimmer and W. M Bayliss, "The Separation of Phosphorus from Caseinogen by the Action of Enzymes and Alkali," *Journal of Physiology (London)* 33(1906): 439–61.

[4] F. Pentimalli and G. Schmidt, "Über das Verhalten der Phosphorfraktionen im Blutplasma sarkomkranker Hühner," *Biochemische Zeitschrift* 282(1935): 62–73.

chemist, a Dr. Jolles, and a Hungarian Jew, a Dr. Kertescz, and found the Italian faculty members very kind. Having purchased a used bicycle, he made Sunday trips into the beautiful countryside, or he took the trolley into downtown Florence and visited museums or admired the doors of the Baptistry. Perhaps best of all, still having his cello, he played chamber music with a young woman, about 20 years of age, and her brother, children of a businessman named Barbecchi, who lived in a house on one of the Florentine hills. He was invited to their home often for dinner, and they played every week. In order to show his gratitude to the young woman, whom he liked, he dared to invite her to go to a concert. She accepted; but when he called for her, her brother went with them. It was quite usual that such a family would protect its young women that way; it would have been "absolutely impossible" for Gerhard, especially as he was a refugee, to develop any relationship with an Italian woman of such a family.

Another most pleasant experience during that year was a five or six-day visit by his mother and his sister Elizabeth, one year his junior, who were able to get temporary exit visas from Germany. Staying in a pensione near the Arno, they were able to explore the great art of Florence, about which his mother was particularly enthusiastic. When they went to see Gerhard's laboratory, Dr. Pentimalli greeted her with a magnificent bouquet of red roses. This would be the last time he saw Elizabeth. She was married to a lawyer, Sepp Grünebaum, who had a significant position, as Oberamtmann, a sort of district or town administrator in Germany, one of the few Jews to have such an appointment. As an Iron Cross recipient for service as a captain in WWI, he still felt safe even in 1934; but he and Elizabeth lived in a small town of about 30,000 people in northern Bavaria and he was prominently visible in such a place, so after the Nazi takeover in 1933, the Bavarian Government moved him to a less exposed post in Munich. During Kristallnacht, the night of the ninth of November 1938, when the synagogues were burned, Jewish-owned stores looted and Jews beaten, arrested, and sometimes killed, in Munich as throughout Germany, a Gestapo car or truck stopped in front of their house. Sepp was arrested and put in a concentration camp, where he was held for six weeks. Early in 1939, soon after he was released, he and Elizabeth emigrated to England. He developed pancreatic cancer and died in the late 1940s. Elizabeth died of thyroid cancer in 1955.

At the time of her visit to Florence, Isabella asked Gerhard, her son, for his advice and consent concerning what to do with the houses that she and her sister co-owned in Nürmberg. Her father, of whom she was very proud, had been a highly respected partial owner and director of a private grammar school for boys in Nürmberg. After the death of her father, in 1917, and later, of her mother, Isabella and her sister, together, owned the boarding school, including a substantial building. Isabella had discussed this problem and her worries about the Nazi regime in detail with Gerhard's uncle Michael, a very well-known lawyer in Nürmberg. He had responded to these concerns saying, "But Isabella, these are all things which are impossible. You have to consider despite of all that happened, 'Wir leben doch in ein Rechtsstaat'—'We still live in a State of Laws.'"

Gerhard, however, had definite opinions about the future behavior of the Nazis, and became impatient and angry when she told him details about the property. He told her curtly, "I'm not interested. I am absolutely sure that nobody of our family will ever get even a penny's worth out of these buildings. You can be absolutely sure that the Nazis will expropriate all the Jewish holdings. You can do what you want; you can figure it out with my lawyer uncle. Now let's enjoy the nice museums and the countryside here. I don't want to hear anything of it. I'll have to earn a living somewhere in the world; but I don't count on anything." Gerhard, having experienced the post-World War I disillusionment, political strife, economic collapse, and the great infla- tion, had little faith at that time in investing for the future. "After having finished with quite some trouble and some financial hardships during my study time, and finally, after having at least made my first step to become a scientist, of becoming a member of the faculty at Frankfurt, now this bum came along and just because of an uneducated demagogue I had to lose out again. So you can imagine I didn't believe in anything."

He did greatly value and enjoy the opportunities he had to live and work in Florence, Naples, and Stockholm, and considered that his pleasure in these pursuits was, in fact, a victory over the Nazis. He focused on making the most of what he could do in the present, certainly with no anticipation of gain from anything in Nazi Germany. Isabella and Elizabeth returned to Germany, where they remained until 1939, as did his youngest sister, Renate, who was about ten years old when Gerhard was in Florence.

Gerhard pointed out that, in 1934, Fascism in Italy did not emphasize racism and was not yet the immediate threat to Jews that Nazism was in Germany.[5] Among the Italian people he met, he did not perceive anti-Semitism. Within a week of Gerhard's arrival, his landlord's older son, a bank employee, invited him to a weekly social evening, with dancing, in the Casa Fascista in Rifredi. It was common under a Fascist regime, including that of the Nazis, that every district had its meeting hall; in Germany, that was where the Brownshirts met. Under the Fascist Party of Italy, there was a big meeting house in Florence and a smaller one in the suburb of Rifredi. Although his

---

[5] The evolution of Mussolini's policies relating to Jews, outlined in this section, is described in M. Michaelis, *Mussolini and the Jews. German-Italian Relations and the Jewish Question in Italy 1922–1945* (Oxford, UK: The Clarendon Press, 1978).

hosts knew that Gerhard was a Jewish refugee from Germany, they found nothing wrong with asking him to come with them to their Fascist meeting house for an evening of dancing. The same comfortable atmosphere prevailed at the university.

In fact, people in Florence and its surroundings still dared to express their witty skepticism well into the years of the Fascist regime. Gerhard recalled that, in the countryside, almost every farm house had to have a big inscription, quoting one of Mussolini's speeches, in letters a meter high so one could see it from a distance. On one farmhouse he saw such a quote saying (translated): "Only the Lord knows how to bend the ardent will of the Fascists. The people and events will never do it"; but under that sign someone had added, "Let's hope in the Lord."

"Of course, you knew that you were in a Fascist country because in the daily newspaper, *ll Matino* in Florence, which had only four pages anyway, three and a half pages were usually devoted to some description of Il Duce inaugurating a new piazza called in memory of this or that Fascist politician. Half of the front page was usually taken up by photographs of Il Duce on a white horse. But otherwise you noticed very little. Everyone told me that it's marvelous how well the trains run on time—that was really true." People also admired large projects, such as the building of towns along the coast south of Rome in drained and reclaimed Pontine marshes.

Two months after his arrival in Florence, Gerhard was shocked by harsh news from Germany, about actions that ominously strengthened Hitler's grip on power, eliminating potential competition even within the Nazi fold. The Italian newspapers carried reports of these actions, known as the "Röhm Putsch" or "The Night of Long Knives"—Hitler's purging of the too-independent forces of the Sturmabteilung (SA), the paramilitary Brownshirt stormtroopers.[6] Between June 30 and July 2, 1934, the Gestapo and the SS executed and/or arrested large numbers of SA members. Their leader, Ernst Röhm, an erstwhile ally of Hitler, was arrested and later shot. Former Chancellor Schleicher was also executed in this purge. Gerhard's pessimism about Germany's future deepened.

In Italy, in 1934, the official policy of Mussolini's Fascist government was anti-racist. Mussolini publicly reassured his Jewish citizens, in an interview with Emil Ludwig in 1932, that anti-Semitism is not part of Italian Fascism, and he criticized Hitler's racism in 1933. Although Hitler and Mussolini had explored warming relationships between their countries, they cooled toward each other in 1934, a year during which there was, in fact, a deepening rift between the two Fascist states. At that time Mussolini was even courting

---

[6] R. J. Evans, "Night of Long Knives," in *The Third Reich in Power* (New York: Penguin Books, 2006).

Western support for his ambitions in North Africa, particularly in Ethiopia (Abyssinia), and he feared Hitler's actions in Austria and the possibility that the Germans would aim to cross the Brenner Pass. He also courted support of Jews, both in Italy and Palestine, and held meetings with national and international Jewish leaders. On the other hand, his government had changed the status of Jews soon after he took power in 1922, by declaring the Catholic religion as the dominant religion of the country, ending the previously established equality of all religious faiths. The Mussolini-controlled Fascist press printed articles critical of international Jewish bankers, Zionists, Masons, and Bolshevists. Mussolini quietly backed anti-Jewish measures, such as blocking certain academic appointments, and even encouraged some German anti-Jewish policies. The Fascist press had been particularly anti-Jewish in 1933. Although 1934 was a period of relaxation of that public antagonism, it was also a period of wariness about mixed signals from government sources.

Near the end of 1934, more clouds appeared in Italy, with increasing tension in Ethiopia (Abyssinia), a confrontation that had grown ever since Italy had built a fort, against League of Nations agreements, in Ual Ual, in the Somali region of Ethiopia, in 1930. In November 1934, there was a confrontation between Ethiopian militia and Italian-serving Somali troops. Attempts to calm the situation continued, to some extent through the League of Nations, for several months into the next year; but in October 1935, Italian forces invaded Ethiopia and soon conquered a much more primitive military force. England and France strongly condemned the invasion; and especially after the League of Nations named Italy the aggressor, Mussolini moved toward alliance with Hitler. By 1938, he developed his own racial anti-Semitism program in Italy. Before 1934, he had already declared racial classifications and restrictions differentiating Italian colonial settlers from indigenous populations in Italian-occupied Northern African and Mediterranean countries.

Gerhard did feel a change in atmosphere during the time he was in Florence. He became concerned, seeing an increasing number of books with racial references. He recalled this as a phenomenon that developed in the span of six weeks, with the appearance of books he had never seen before. "I mean, show windows which formerly were books of art and guides to Florence suddenly were full with racial literature, and I was worried."

He was so much in love with Florence that he would have been eager to stay there. The changing atmosphere, however, made him pay attention to other options. In May 1934, he learned from the Society for the Protection of Science and Learning (SPSL) in England that the potential position in Jerusalem they were exploring for him was not available. Another opportunity, however, arose in mid-June, in the form of a letter from the Carnegie Corporation in New York, informing him that the corporation was granting some funds to Queen's University in Kingston, Ontario, Canada, and asking him whether he would be interested in a two-year research fellowship at that university. As Gerhard had not applied to the Carnegie Corporation, he concluded that they must have obtained his name and address from the SPSL. He turned for advice to Pentimalli, who responded that he was very sorry to hear such news, but much as he would like Gerhard to work with him for a longer time, the wise thing to do would be to accept this position because this is a new era and nobody knows what will come of it. Thus, just a little more than a year after he left Germany, having worked in temporary positions in Naples, Stockholm, and Florence, Gerhard was contemplating another move to another temporary position, this time across the Atlantic Ocean.

Nine

Crossing to the New World: Queen's University

AID FOR REFUGEE SCHOLARS AND PHYSICIANS

In parallel with formation of the Society for the Protection of Science and Learning (SPSL) in England, in response to the academic displacements in Germany resulting from the March law to "purify" the German Civil Service, organizations to aid scholars and professionals who had to flee Germany were formed in the United States as well. The Emergency Committee in Aid of Displaced Foreign Scholars was formed when Dr. Alfred E. Cohn, a distinguished scientist at the Rockefeller Institute, sought advice from the Institute of International Education, an organization that already had experience in placing foreign students. Dr. Stephen Duggan, Director of that Institute, responded by joining the new effort.[1] He and Dr. Cohn, along with philanthropists Bernard Flexner and Fred M. Stein, then recruited officers and a General Committee, with 23 university representatives, was formed by mid-June 1933. The Committee, with funds from Jewish philanthropists, including Mr. Flexner, Mr. Stein, and Felix Warburg, and from the New York Foundation, eventually helped to place 335 scholars (from among 6,000 who appealed to the Committee) in many fields of learning, by providing funds to universities and foundations for temporary fellowships. It focused mainly on the humanities, leaving aid for physicians and lawyers to other organizations. During its operation from 1933 to 1945, this Committee for Displaced Scholars played an important role in saving the lives and, in the best circumstances, recreating the careers of experts in language and language teaching, art historians, archaeologists, philosophers, musicians and musicologists, economists, social scientists, and some natural scientists. Gerhard was on the list of those who did not receive a grant from this Committee; and so was the eminent physician scientist, Dr.

---

[1] S. Duggan and B. Drury, *The Rescue of Science and Learning. The Story of the Emergency Committee in Aid of Displaced Foreign Scholars* (New York: MacMillan, 1948). For the records of this organization, see: Archives of the Emergency Committee In Aid of Displaced Foreign Scholars Records 1927–1949 MssCol 922, New York Public Library, Humanities and Social Sciences Library, Manuscripts and Archives Division. Online records accessed August 15, 2014, at http://archives.nypl.org/mss/922.

Siegfried Thannhauser, who would nevertheless come to America and play an important role in Gerhard's life and career.

A second and related effort, which did find a place for Gerhard, was the Emergency Committee in Aid of Displaced Foreign Physicians. Growing out of the original Committee for Scholars, again at the initiative of Dr. Alfred E. Cohn, it was formed in November 1933, with Dr. Bernard Sachs as Chairman and Dr. George Baehr as Secretary. It encountered significant problems trying to place clinicians during the time of the Great Depression. Practicing doctors, especially in the cities where refugees were growing in number, objected to the idea of helping potential competitors establish new practices during these economically difficult times. Barriers were set for certification and entry of refugees into clinical positions. The Committee, therefore, emphasized placement of research physicians, such as Gerhard, into academic positions and tried to draw clinicians to places outside New York. The Committee for Displaced Physicians sent notices to universities across the continent and announced its mission in the *Journal of the American Medical Association*,[2] seeking positions for which it or the Carnegie Corporation would provide temporary fellowship grants—to the institutions—to support salaries of refugee physician scientists. The Committee funds were, like those of the Committee for Displaced Scholars, obtained from philanthropists and the New York Foundation. As some of the Carnegie Corporation funds were available on the condition that they would be spent in the British Dominions and Colonies, Canadian universities were invited to participate.

These notices led to a May 10, 1934, response from Queen's University in Kingston, Ontario, saying, "We can provide [research] facilities for a biochemist, and should welcome here one of several refugees whose work would be of great advantage to a very important department in which we are at present weak." After an exchange of letters among the Principal of the University, the Emergency Committee, and the Carnegie Corporation, a fellowship grant was approved by late May, with half of the money to come from the Committee funds and half from the Carnegie Corporation.[3] This was one of only six positions provided for displaced German academics among all the universities in Canada.[4] Gerhard received a letter from Dr. Baehr and, at about the same time, a formal invitation from the University, dated June 15, 1934. It offered a stipend of $2,500 per year for two years, a very large sum for Gerhard in those days. He would be in the Department of Chemistry, which served the Faculties of Arts and Science, Medicine, and Engineering.

---

[2] G. Baehr, "Emergency Committee in Aid of Displaced Foreign Physicians," *Journal of the American Medical Association* 101(1933): 1900.

[3] Archives of Queen's University, Kingston, ON, Canada.

[4] D. Zimmerman, "Narrow-Minded People: Canadian Universities and the Academic Refugee Crises, 1933–1941" *Canadian Historical Review* 88(2007): 291–315.

The university would make facilities available but there would be no additional funding for special supplies. After his chat with Dr. Pentimalli in Florence, Gerhard responded by the end of June that he would be glad to take up the offer, but wanted to complete some of his current work and begin in Kingston late that year or early in the next year, a condition accepted by the university.

He was ready to leave Florence February 22, 1935, planning to reach New York by the 12th of March, on his way to Kingston. There was a delay, partly to complete visa arrangements with the Canadian Department of Immigration and Colonization. Eventually, Gerhard boarded a boat, a relatively small liner, in Trieste, "mainly because it was supposed to stop in Greece (in Patras) and Sicily and Naples—and then in the Azores and finally New York. I had saved enough from my fellowship to pay third-class transportation. It was enjoyable." Unfortunately, the stop in Greece did not materialize because there had been an attempted coup against the government March 1,[5] the day before the ship arrived there; but he did enjoy other stops, reached New York March 25, and telegraphed Dr. Fyfe, Principal of Queen's University to say he would be in Kingston soon.

In the days he spent in New York before going on to Kingston, he went to the Rockefeller Institute to visit Dr. Phoebus Aaron Levene, widely known simply as P. A. Levene, to whom he had been introduced by Embden in 1929 on a visit to New York after the International Physiology Congress in Boston. It was to Levene that Embden had turned, requesting an authentic sample of adenylic acid, when Gerhard had found the different behavior of adenylic acid samples that he had prepared from two different sources. Levene (1869–1940), a senior member at the Rockefeller Institute, was one of the world's highest authorities on the chemistry of nucleic acids.[6] Born in Lithuania, he grew up and studied medicine in St. Petersburg, Russia. In response to an environment of pogroms, his family emigrated to the United States in 1893. Levene studied at Columbia University, practiced medicine, and became involved in biochemical research, at first in the chemical structure of sugars. He became head of the biochemical laboratory at the Rockefeller Institute, where he stayed for the rest of his career, and where he discovered: 2-deoxyribose as the sugar of DNA, nucleotides (i.e., base–sugar–phosphate) as the unit structures of DNA, and the fact that the nucleotides are linked together by the phosphate groups. About 1910 he formulated the hypothesis that a chain of four nucleotide units—a tetranucleotide—was the fundamental structure of DNA. As DNA was not known, at that time, to be the carrier of

---

[5] Supporters within the Greek military of former Prime Minister Eleftherios Venizelos launched an unsuccessful coup attempt against the government.

[6] D. D. Van Slyke and W. A. Jacobs, "Biographical Memoir of Phoebus Aaron Theodor Levene 1869–1940," *National Academy of Sciences of the United States of America Biographical Memoirs*, 23(1943): 75–126. Available online at http://www.nasonline.org/publications/biographical-memoirs/memoir-pdfs/levene-phoebus-a.pdf.

genetic information, the fact that the proposed structure was too simple for such a function was not a barrier to its wide acceptance. The hypothesis[7] was still standing when Gerhard and Levene met at the Rockefeller Institute. Neither of them knew then that Gerhard would return in two years to work in Levene's laboratory or that Gerhard's research results would play a role in the overthrow of the tetranucleotide hypothesis.

Speaking of this 1935 visit with Levene, Gerhard recalled that "He gave me very good advice. In a worried manner, he said, 'Now, I would suggest not to make too many demands, where you go, with regard to equipment, or about having to spend some time for teaching. Do everything they want you to do and the main thing, please, do not spoil their good will by showing anything of a conceitedness which is often shown by Germans who are coming through here.'" He mentioned, as an example, a particular biochemist from Berlin, quite well known, who created great difficulties for himself by not being satisfied with what he found at his new home in America. Ironically, and probably unknown to Gerhard, that difficult biochemist had been a candidate for the Queen's University position before Gerhard but had chosen to settle in a different place.

Levene pointed out that great efforts have been made in America, especially considering it was a time of the great economic depression, to place as many displaced German scholars as possible. Levene and the Rockefeller Institute were, in fact, important players in those efforts. Of course, he pointed out, it would benefit certain smaller American colleges that were not principally interested in research that, through these fellowships, they could get very experienced investigators on their staff, and they might have stimulating effects on the academic activities of these colleges. But, he said, "One has to be aware of the fact that a number of these places are not equipped with the most modern instruments, the most expensive instruments and one should not immediately expect these colleges to put at the disposal of refugee scientists splendid laboratories like those in the Kaiser Wilhelm Institutes (now the Max Planck Institutes)."

Scholars from these German institutions did face significant adjustments, in language, research facilities, teaching requirements, and relationships among faculty and students and administration when they were placed in American colleges. The efforts and, mainly, the success of scholars in many fields in adapting to America has been described in some detail by Dr. Stephen Duggan and Betty Drury, Chairman and Executive Secretary of the Emergency Committee for Displaced Foreign Scholars. Language was certainly

---

[7]P. A. Levene and W. A. Jacobs, "On The Structure of Thymus Nucleic Acid," *Journal of Biological Chemistry* 12(1912): 411; P. A. Levene and E. S. London, "The Structure of Thymonucleic Acid," *Journal of Biological Chemistry* 83(1929): 793–802.

a factor for Gerhard. His gymnasium education included just one year of English, mostly very formal. His first letters to Queen's University were in German, and it was in that language that he and Levene were able to converse most easily. He undoubtedly had learned to read English scientific articles, and he did become proficient in his new language. The later letters to the university, still before his arrival, were in English.

Levene facilitated another personal contact in New York for Gerhard, who otherwise had no relatives or friends in the city. Levene suggested that he should call Dr. Harry Sobotka, who had studied with Willstätter in Munich, had been at the Rockefeller Institute, and was by then head of the laboratory at Mt. Sinai Hospital. Dr. Sobotka had married a professional harpsichordist, Vienna-born Yella Pessl, who had immigrated to the United States in 1932. Gerhard did call and spent a fine evening with them. As in all his major moves, he had carried his cello with him. That evening he played sonatas together with Ms Pessl.

Traveling by train from New York, Gerhard reached Kingston on Tuesday, April 2. The next morning he received a phone call from a reporter for the *Kingston Whig Standard* who wanted to interview him; arrival of a refugee scientist from Germany was a significant news item in this city of about 22,000 people.

Gerhard was welcomed warmly by Professor Arthur C. Neish,[8] head of the Department of Chemistry at Queen's. The chemistry faculty also included Professor J. A. McRae, an organic chemist; Associate Professor J. F. Logan, a biochemist; three assistant professors and one instructor. With the fellowship, Gerhard's role would be strictly in research. For the first time since Frankfurt and his brief stay in Naples, he would be independent in his choice of subject and in the conduct of research. He felt well provided for in laboratory space and supplies. With an approval signature by Dr. Neish or Logan, he was able to purchase all the additional equipment and materials he needed.

NUCLEOPROTEINS, PHOSPHOPROTEINS, AND EMBRYOS

For his research topic, he drew together the various threads that led him to nucleic acids and nucleoprotein: his original work with the nucleotides adenylic acid and guanylic acid and their metabolism; his finding in Naples that virtually all the purines in sea urchin eggs were bound in compounds that

---

[8]Prof. Arthur C. Neish (1877–1948) a native of Kingston, ON, earned his BA at Queen's University, MA and PhD at Colubmia, and joined the Queen's faculty in 1901. He rose to become Professor and Chairman, Department of Chemistry, 1919–41; E. J. Lemieux and J.-F. Legault, "The A. C. Neish Competition 1947–1997," *Canadian Chemical News* 49(1997): 36.

were large (high molecular weight) and acid-insoluble; his 1929 visit with P. A. Levene, who showed him and Embden crystals of the recently discovered DNA sugar, 2-deoxyribose; his very preliminary work, in Stockholm, on fractionation of nucleoprotein from fish sperm; and his growing interest in the susceptibility of different kinds of phosphate linkages to chemical or enzymatic hydrolysis. He had also been impressed by Robert Feulgen's 1923 monograph "The Nucleic Acids,"[9] which he consulted extensively in his work on nucleotides in Frankfurt, and an earlier monograph on nucleic acids written by Walter Jones of Johns Hopkins University.[10] One of the many unresolved issues was how nucleic acid and protein are linked in the nuclear nucleohistone or nucleoprotamine. Are they joined through chemical covalent bonds or, as had been suggested by Feulgen, by ionic interactions, as the negatively and positively charges components of a salt-like compound? While still in Frankfurt, Gerhard had tried but was unable to separate the nucleic acid and protein components of nucleoprotein by electrophoresis, so he leaned toward the idea of covalent bonding—an idea he would have to discard.

Choosing to study nucleoprotein, he needed a source of nucleus-rich organs, such as thymus or spleen. Here his Jewish identity served him well. As a Jewish refugee, he received a warm welcome from the Jewish community in Kingston, and the rabbi was particularly kind to him. The rabbi was also the supervisor of kosher slaughter, and introduced him to the people at the commercial slaughterhouse, where he could get calf thymus and spleen, with never a charge being levied.

Gerhard decided to approach some questions of nucleic acid structure by studying their depolymerization into nucleotides and release of phosphate from the nucleotides. Ribonucleic acid (RNA) could be broken down chemically by exposure to alkali, providing a major source of nucleotides; but, as Gerhard had shown in Frankfurt, nucleotides formed this way, adenylic acid, for example, had phosphate linked at a different position than naturally occurring adenylic acid from muscle tissue. In the case of DNA, which was stable in alkali and released free purines but not nucleotides in acid, a different approach—enzymatic depolymerization—was needed.

Enzymes capable of degrading DNA were found in the intestine, and they were used by Levene in his discovery of deoxyribose as the DNA sugar. Levene did not, however, have purified enzyme. Rather, in his first successful experiments, he used introduced DNA into a surgically isolated closed loop of dog intestine, kept alive under the skin of living animals; after two hours of incubation, he recovered the degraded DNA products from intestinal fluid;

---

[9] R. J. Feulgen, *Chemie und Physiologie der Nucleinstoff* (Berlin: Born-traeger, 1923).

[10] W. Jones, *Nucleic Acids. Their Chemical Properties and Physiological Conduct* (London: Longman Green, 1914). This was part of a series of monographs on biochemistry.

and he repeated this injection and collection daily over several weeks. As intestine has both *deoxyribonuclease* and a very active *monophosphatase*, most of the products were nucleosides, lacking phosphate, rather than nucleotides. Levene had crystallized the nucleosides, degraded the purine nucleosides with dilute hydrochloric acid, and then recovered the sugar and purines separately. A simpler source of intestinal enzymes was found in fluid samples taken from patients with gastrointestinal ulcers or extracts of intestinal tissue; and a very active partially purified *intestinal phosphatase* could be prepared as a glycerol extract of intestinal mucosal tissue with a procedure described by Siegfried Thannhauser. The *phosphatase*, bound in the intestinal cell wall structure, was not soluble and could not be completely purified, but the very active partially purified enzyme would become a major research tool. Gerhard's research approach was to compare the susceptibility of phosphate groups in nucleic acid and nucleohistone to cleavage by *intestinal monophosphatase*, an enzyme that cleaves phosphate of a single nucleotide or phosphate at the end of a chain of nucleotides, but does not cleave the phosphate linkages between nucleotides within the chain. With protein-free DNA, all the end-of-chain phosphate was cleaved during a 24-hour incubation with enzyme. With nucleohistone, however, only about 20% of the release occurred in 24 hours, and the further cleavage occurred only very slowly. He concluded that about 20% of the DNA in the nucleoprotein preparation was actually free, unbound, nucleic acid, and that in the real nucleoprotein complex, the nucleic acid–protein link was preventing access of the enzyme to the DNA. He then used a pancreatic extract, known as *pancreatin,* as a source of *trypsin* to break down the protein component of nucleohistone and found that, after such treatment, the DNA phosphate was cleaved at a much faster rate, the same rate as in free DNA. He submitted a short note on this finding to the journal *Science,* an article which he thought was probably reviewed by P. A. Levene. It was accepted.[11]

In the original manuscript sent to *Science,* he suggested that the increased accessibility of phosphate groups after *trypsin* treatment was evidence for covalent DNA-protein linkage, which may have been cleaved by the *trypsin.* Just after the note was accepted for *Science,* however, he discovered a newly published article by Feulgen that caused him concern.[12] Feulgen made his DNA preparation by treating nucleohistone with alkali, which degraded the protein but left the DNA relatively intact, as a very viscous gel, indicating it was a very large molecule. Feulgen showed that pancreatic extract contains

---

[11] G. Schmidt, "A Chemical Difference between Protein-Linked and Free Nucleic Acid," *Science* 83(1936): 15.

[12] R. Feulgen, "Über nucleo-*gelase,*" *Zeitschrift für physiologische Chemie* 237(1935): 261.

not only the protein-degrading enzyme *trypsin*, but also an enzyme that elimi-
nated the viscosity of the DNA gel by degrading it into small molecular
fragments, an enzyme that came to be known as *deoxyribonuclease* (*DNase*).
Interpretation of Gerhard's results was thus clouded by the very strong likeli-
hood that, quite apart from altering any DNA-protein linkage, his pancreatic
extract had broken the DNA into short oligonucleotides, an effect which, in
itself, would increase the number of end-of-chain phosphates susceptible to
monophosphatase action. With that insight, he modified the conclusions of
his *Science* letter before it was published. The printed version stated that
the difference in behavior between free nucleic acid and nucleoprotein was
"produced by the linkage of the nucleic acid component with the protein,"
but did not comment on the nature of the linkage.

The details of enzyme, DNA, and nucleoprotein preparation, and the quantitative results of the experiments were published in 1936 in a regular article
in the first volume of a new journal, *Enzymologia* (which became *Molecular
and Cellular Biochemistry* in 1973).[13]

These studies clearly showed that DNA in nucleohistone was substantially
protected from hydrolytic activity of his preparation of *intestinal phosphatase*,
and that incubation of nucleohistone at 70°C did not increase accessibility
of DNA to digestion. Whereas use of a crude pancreatic extract as a source
of *trypsin* did increase accessibility, use of a purified *trypsin* did not; that is,
both nucleic acid and protein components were protected when they were
bound together in the nucleoprotein complex. The phosphate of free DNA
was much more accessible than that of nucleoprotein. Although Gerhard
referred to Feulgen's recent work as suggesting that depolymerization of
DNA was a step that would precede dephosphorylation, and considered it a
possibility, his own analyses of the amount of phosphate and free purine
released from free DNA still seemed to fit the idea of DNA as a tetranucleotide,
which served as the basis for his calculations.

A quite different line of research, reflecting his continuing interest in
embryonic development, also developed in his years at Queen's, leading to
a publication in *Enzymologia* in 1937.[14] He had been impressed by a 1935
report, by H. Vollmar, that chicken egg embryos could be sustained outside
the shell, incubated in egg white, for 10 to 16 days. He decided to test the
importance of egg white itself in that phenomenon. Egg white is eventually
taken up by the developing embryo, but it was not known whether it served
as a substrate for energy-producing metabolism or whether it contained specific

---

[13] G. Schmidt, "Effect of Nucleophosphatase upon Thymus Nucleohistone: Effect of Enzymes on Proteins
with Prosthetic Groups I," *Enzymologia* 1(1937): 135–41.
[14] G. Schmidt, "Growth-Stimulating Effect of Egg White and Its Importance for Embryonic Development,"
Enzymologia 4(1937): 40–48.

growth and development-promoting chemicals. He developed a method for isolating the intact yolk with the embryonic disc on an exposed surface, and for replacing the medium surrounding it with either isotonic saline or solutions containing varying dilutions of egg white. Based on experiments with this preparation,[15] he concluded that some organic component of the egg white, something more than glucose, was indeed required for normal growth and development of the embryo.

The *Science* note on nucleohistone and the articles in *Enzymologia* were presented in English, with very few spelling, grammatical, or stylistic indicators that this was not the writer's native language. He presented his results at the first meeting of the Canadian Physiological Society in Toronto on October 19, 1935. His communication, "The Effect of Nucleophosphatase Upon Nucleohistone" was one of eight scientific presentations at that meeting. Gerhard was also invited twice to present seminars at the University of Toronto, one of them on the binding of histones to DNA. These invitations came from Professor Laurence Irving, who had previously worked in Embden's lab, and who was, by 1935, a professor of biology in Toronto.[16]

ADAPTATION TO SOCIAL LIFE IN KINGSTON

When Gerhard first arrived in Kingston, the university had arranged temporary accommodation for him in rooms of the Student Union, an organization overseen by a very helpful Colonel MacDonald, who also personally provided him with transportation. After a few months he found a room in a private home. Though he found the cold winter, with its abundant snow, very difficult, he also found, and was included in, a lively social community. There were many discussion clubs and musical groups and amateur theater, though few

---

[15] When egg white was diluted with one-third or one-fourth of its volume of buffered saline, it sustained growth for 4 to 6 days, but then uptake of fluid into the yolk space led to rupture of the vitelline membranes and death of the embryo. There was no growth of the blastoderm disc in buffered salt solutions without egg white; in particular, the presence of potassium was toxic. With most salt solutions, the damage was irreversible; only with Ringer-bicarbonate was it possible to restart growth by the subsequent addition of egg white. Addition of egg white to comprise between 5% and 20% of the total fluid did sustain growth in comparison with the Ringer-bicarbonate alone. He concluded that organic components of the egg white were responsible. Attempting to learn which components were required, he used dialysis against Ringer-bicarbonate solution to obtain the egg white proteins devoid of small molecules. The protein alone did not support embryo growth. Neither did the dialyzate. When the two were recombined, however, the growth-promoting effect was reconstituted. Addition of glucose, in the same concentration as in original egg white had some effect, but not as much as egg white itself.

[16] A short biography of Laurence Irving, including reference to his work with Embden: W. R. Dawson, "Laurence Irving. An Appreciation," *Physiological and Biochemical Zoology* 80(2007): 9–24. Irving became known for his investigations of the physiology of diving mammals, the respiratory properties of fish blood, and cold adaptation and acclimatization in poikilotherms and homeotherms, including man. He moved to Swarthmore University and later to Alaska, becoming the first scientific director of the Arctic Research Laboratory at Barrow, Alaska, and as the founding director of the Institute of Arctic Biology at the University of Alaska, Fairbanks.

performances by well-known musicians or theaters from outside Kingston. Many of Gerhard's warm associations were with other faculty members, particularly the plant physiologist Gleb Krotkov, originally from Prague, who had earned his PhD at the University of Toronto in 1934 and had just joined the Queen's University faculty. He also met the geologist Harold Fairbairn, who also played the cello, and who went on to join the faculty of Massachusetts Institute of Technology. He made other musical associations with a professor of music and a mathematician.

The highlights of cultural associations, however, were the musical evenings held by the professor of pharmacology, Thomas Gibson. Gibson, educated through medical school in Edinburgh, Scotland, had come to Canada in 1895, served as physician for four Governors-General, and then practiced medicine in Ottawa before he was recruited to head the Department of Pharmacology and Therapeutics at Queen's. In his youth he had also shown a remarkable musical talent as a performing pianist, a skill he continued to express in Kingston. Gerhard recalls him as an outstanding pianist who had performed on radio broadcasts. Gibson and his wife were very sociable people, involved in the Society of St. Andrews, and he arranged for Gerhard to be invited to the annual dinner of the Society. He also invited Gerhard to his home for social events, one of which required formal dress. Gerhard had brought a tuxedo with him from Germany—probably the one he had bought for the party celebrating his faculty appointment at Frankfurt—and a bow tie that was not pre-tied; his father would never allow him to use a pre-tied bow, but he never became really proficient at tying it himself. When he arrived at the Gibson home, the Professor took him into the wardrobe and personally fixed the tie, and then introduced him to Mrs. Gibson and guests—a ritual that was repeated on other occasions. Eventually Gerhard was invited almost weekly to the musical evenings.

It was another invitation that led to a legendary feat of Gerhard's two years in Kingston. On weekends he was a very active person, but usually on his own. He purchased a bicycle and would go for long rides, once even cycling to Ottawa, about 125 miles away. One spring/summer Sunday in '36, he started on a bicycle ride but turned back toward Kingston with the thought of taking up an invitation given by the sociable wife of one of the faculty members. The previous autumn, she had told him that she and her husband have a summer home on Garden Island and that he would be welcome to come and visit them when he had a chance. He parked the bike, left his belongings with a guard at a bathhouse, entered the water and started to swim toward Garden Island, about 1.7 miles into Lake Ontario, south of Kingston, opposite the lakeside of Queen's University. Having made it all the way, he came up on an island beach in the afternoon. He assumed he could find someone on the beach, maybe even some other faculty members, who would know the whereabouts of the home of Dr. G. There were very few people in sight, but eventually a young boy came along and showed him the way to the "blue cottage," as Dr. G.'s summer home was known. He came to the house and knocked on the door. Mrs. G. answered. She was dumbfounded, speechless, seeing him fresh from the water in just his bathing suit, especially as, just at that time, she and Dr. G. were hosting a cocktail party for the whole science faculty. There was a flurry of activity. The professor of physics, who also had a summer home, brought some trousers and someone brought a shirt and soon Gerhard was able to join the party, where, amid laughter—and some embarrassment for the host—his feat was greatly admired. After not too long a stay he said it was time to go back because it would take another hour and a half to two hours to swim and he wanted to return to Kingston before dark. His hosts insisted that he would not swim back. Somewhat later, they took him to the dock for the ferry that made the trip back to Kingston. When he got back to the bathhouse, he found the police had been searching for him because, after five o'clock, the bathhouse guard, who still had his clothes and wallet, had reported him as missing.

As a foreign guest, Gerhard met other exchange students from abroad. With one, a student from France, he became close friends. Meetings with another one, who came from Germany, were much less comfortable. The chairman of the Department of German, Professor Heinrich Henel, originally from Frankfurt himself, arranged events to which he invited both Gerhard and a German exchange student. That student was very likely a Nazi Party member or, at least, having been selected for such an exchange program in 1935, must have been approved by the Party. While Henel brought them together with the sense that "it is so nice for these compatriots to meet," Gerhard felt the irony of this imposed social contact between "the Nazi and the Jew."

One of the few musical recitals given by someone from outside Kingston, probably in 1936, was a memorable one for Gerhard. It was an oratorio presentation by Magda Spiegel, a Jewish contralto from the Frankfurt Opera, whom he had heard in performance when he was still in Frankfurt. He was moved by the recital and introduced himself to her after the concert; he was shaken when he learned of the very difficult conditions for her family in Germany from 1935 on. Poignantly, he recalled that much later, in 1964, when the University of Frankfurt was celebrating its 50th anniversary, it sent him a commemorative booklet, which included the history of the university. It mentioned an event in early 1942, when a senior university faculty member, the dermatologist Karl Herxheimer, was being loaded on a freight train for deportation to "the east," where he died in Theresienstadt. The singer Magda

Spiegel was also among the victims on that train. She had been kicked and had fallen, and was helped up by the aged Herxheimer. She was on the way to her death in Auschwitz.

Life in Germany had become more difficult for Gerhard's family as well, especially after promulgation of the Nürmberg laws in 1935. He received a letter from his mother, telling him what had happened to the properties she had discussed with him when they met in Italy. She had received notification to come to the Gestapo office in Nürmberg. Because her sister had died earlier and her brother, who was co-owner of these houses, also died, she was the only owner remaining in Germany. She was forced to sell these houses at a token value. In other ways as well, Jews were officially declared second-class citizens. They had to wear the yellow star; and each male Jew had to have in his passport the middle name Isaac, and each woman had to have the middle name Sarah, to show to everyone—to every policeman—that he or she was Jewish. Furthermore, his mother could no longer employ a German household helper. She had to move to a different section of the city. Gerhard realized that the changes in Germany were not short-term events that would soon pass; he had to plan to help his mother and sisters get out.

WHAT WILL FOLLOW THE TWO-YEAR FELLOWSHIP?

Gerhard's fellowship at Queen's was set to expire in April 1937. As that time approached, he hoped there would be a way for him to stay at the university. His hope ran up against some administrative changes and uncertainties. The principal, Dr. Fyfe, who had brought Gerhard to Queen's, left the university in 1936 in order to take up the position of principal and vice-chancellor of the University of Aberdeen. His successor was Professor Robert C. Wallace, originally from Scotland, who had come to Canada to head the Department of Geology at the University of Manitoba and then went on to become the second president of the University of Alberta in 1928.

Some hopeful news for Gerhard was the university's plan to create a Department of Biochemistry in 1937; up until that time, the subject was taught to medical students by a member of the Department of Chemistry, Professor Logan. Gerhard was even asked for advice on where the new department should be located; he recommended it should be near the basic medical science departments rather than the clinical departments. One of the candidates for chairman of the new department was E. W. McHenry, then at the Connaught Laboratories in Toronto, studying preparation of liver extracts for treatment of pernicious anemia. McHenry would have been interested in retaining Gerhard on his staff. The recruitment and negotiations were prolonged and eventually the chairman's position was offered to Robert Sinclair, who had been a student of the widely known lipid biochemist Walter R. Bloor in Rochester, New York.

By the time the selection of a chairman was settled, it was too late for Gerhard, who had been compelled to look for an alternative. Along the way, he had made an appointment to talk with Vice-Principal McNeill, hoping McNeill would offer a way to stay until the new department was formed. Instead, McNeill was enthusiastic about another option, one that had been offered to Gerhard but did not interest him. That proposal was to join a clinical study begun by Hendry C. Connell, a specialist in eye, ear, nose, and throat disease, who believed he had an effective new treatment for cancer. Hoping to direct bacterial protein-degrading enzymes specifically toward destruction of cancer cells, he incubated cultures of *B. histolyticus* with pieces of cancer tissue and then prepared a sterile filtrate of the culture fluid, which did contain proteolytic enzymes. The fluid, which was given the name "Ensol," was administered to patients with "incurable" late-stage cancer disease and Connell reported some remarkable responses in an article for the *Canadian Medical Association Journal*. These reports led to a rush of potential recipients to Kingston, enough to affect housing availability in the city, and attracted funds for pursuit of the research. Additional hospital space was made available. Gerhard, on the other hand, was increasingly interested in research on nucleic acids and cell division, was not convinced of the fundamental rationale for Ensol use. He had seen other examples, in Frankfurt, of clinicians devising new, and eventually fruitless, treatments based on poor fundamental scientific understanding. Later, he was glad he did not join the Ensol study, because, in the end, it did not prove effective. Worse, as he read in *The New York Times* in 1938, a Philadelphia lab licensed to produce an Ensol equivalent distributed a product that caused several deaths in the United States, particularly in Florida, due to the presence of tetanus toxin in the filtrate fluid.[17] A skeptical article in *Time Magazine* in 1938 questioned the whole study and evaluation of the treatment.[18]

Because he could not wait for a decision to be made about who would head the new Department of Biochemistry, Gerhard wrote a letter to P. A. Levene in New York, as he had done when he was leaving Germany for Italy in the spring of 1933. This time, to his great joy, the response was positive; he was invited to take a position as an assistant in Levene's department at the Rockefeller Institute, to begin July 1, 1937.

---

[17] "Discoverer of Drug is Ready to Aid: The Laboratory of Dr. Hendry Connell, Where 'Ensol' Is Manufactured, Will Be Inspected by Officials of the Federal Department of Health, "*New York Times*, March 31, 1938; "More Fatal Serum Is Found in Florida," *New York Times*, April 3, 1938; "Disavows Cancer Cure," *New York Times*, October 13, 1938.

[18] "Medicine: Ensol for Cancer," *Time Magazine* Monday, Oct. 14, 1938.

Ten

Joy Once More in Italy:
A Fateful Interlude

T he joy of knowing his next step would be with one of the great scientists in a renowned research institute was accompanied by what seemed at first to be only a wishful idea of how to use the two months of May and June, between the end of his fellowship in Kingston and the beginning of his work in New York. Could he possibly get to Italy during that time, perhaps to meet his mother there once more? His salary at Queen's had been $2500 per year and he had lived frugally because he never lost sight of the need to bring his mother and sister over to America.

He learned it could be done. Contacting the Italian Line office in Toronto, he was treated very warmly and was told he could get a third-class ticket on an Italian steamer from New York for $100 each way; in fact, they would upgrade him to a private cabin at no extra cost. Perhaps, in that Depression era, few people were traveling. In the end, he was given a cabin with four beds all to himself.

There was one more problem to solve, both for travel and for his move to New York. His German passport, which he had received in 1931, had to be extended beyond 1937. Fortunately, the German Consul in New York did extend it, in spite of the date and Gerhard's Jewish identity. While in Canada he also obtained a student visa for his planned stay in New York for training with Levene; later, he realized that this choice, even though it was easy to achieve, was a bad one. It would not have been that much harder to obtain an immigrant visa, which he would eventually need in order to stay in the United States permanently. He did know, at that time, that he would not return to Germany.

Leaving books and bike behind, he traveled on to New York with his suitcase, knapsack, and cello and boarded the steamer The *Conte de Savoia*, a liner that had been built in 1931 and had its maiden voyage to New York in 1932. He had been able to contact his mother and they agreed to meet in Perugia, capital of the Umbria region in central Italy, 180 km north of Rome. The *Conte de Savoia* docked at Naples. He called an Italian couple who had been his closest friends when he was first in Naples in 1933 and was invited for the night "providing you finish that story you were telling us."

The next day, he traveled by train to Perugia. From the station in a valley, it was a steep climb by trolley to the city center. On the way he noticed an ad for a "modern inn," a prime feature being that one could not smell the kitchen from the rooms of the inn. He did not take that one, but did find a beautiful room on the seventh floor of an older home, a chamber under the roof, with a beautiful view over the Tiber Valley and across it to Assisi. The landlord's family was very welcoming; on one occasion, one of the young men took Gerhard on a tour, riding in the sidecar of a motorbike, across the Appenines, to the Adriatic coast.

He was delighted to be back in Italy, especially in a center of culture, with treasures of Etruscan and early Renaissance art and fountains. He planned to stay in Perugia for a few weeks. One of his associates at the University of Florence two years earlier, who had become a professor of general pathology at the University of Perugia, invited Gerhard to do some paid work in the lab, performing nitrogen determinations related to his research on *hyaluronidase*.

Having found his place to live, Gerhard greeted the next morning in an espresso bar, where it became his habit to take breakfast, and met his mother at her hotel. At the espresso bar he also noticed two young women, both the first morning and then again that evening and the next morning. Hearing them speaking German, he felt very frustrated because "I thought they are certainly Nazis and one could not even talk to a German girl anymore."

Two days later he and his mother went to the neighboring town of Gubbio, famous for its Roman amphitheater, to observe a festival that includes a competition, the Corsa dei Ceri, which takes place every year on May 15, the day dedicated to the patron saint of the town. Three Ceri, wooden structures weighing 400 kilos with statues—Sant'Ubaldo, San Giorgio, and Sant'Antonio—are carried by runners on a 4-km uphill course through the streets up to the Basilica of Sant'Ubaldo, the patron saint. The "race" begins at 6 p.m.; winners are determined on the basis of quality of the run rather than the time it took them to finish the course.

The festival, as always, drew a large crowd. Among the throng of visitors were the two young women Gerhard had seen at breakfast, Edith Straus-Horkheimer and Lilo Erlanger. They had made the trip from Perugia to Gubbio in an unusual way. Having wagered some fellow students that they could get to Gubbio and back without cost, they had hitched a ride in a horse-drawn wagon. On the way to Gubbio, they were drenched in a downpour. On arrival, they changed into what dry clothes they had in their packs—pajamas. Embarrassed at their dress, they separated from the group of art students they were supposed to meet and sat down at an ice cream parlor to plan their strategy. There they noted, at a nearby table, a gentleman wearing a "handsome

checkered jacket," an elderly lady, and a girl of about 12, talking German.
It was Gerhard, with his mother and his sister Renate.

Edith and Lilo saw the lady and girl again, a little later in the day, at a
latteria. Recalling the gentleman with them, they wondered whether someone
who owned such a handsome sport jacket might also own a car and could
give them a ride back to Perugia. They began a conversation and learned
that the Schmidt family had come to Gubbio by bus. They also learned that
they had a connection: the cook for Lilo Erlanger's family had been employed
by Gerhard's sister in Munich. Although Gerhard could not provide a car
ride back, this friendly exchange in Gubbio, his first real encounter with
Edith Straus-Horkheimer, was a fateful meeting. They saw each other again
the next day, back in Perugia, when Edith and one of her students were
walking by the espresso bar. He introduced himself and asked whether he
could accompany her home. They exchanged their stories and saw each other
more during his brief stay in Perugia. Edith recalled that he asked her if she
knew someone who could type a German scientific manuscript. She did it;
and, in turn, he corrected a paper on Galileo she had written in Italian, in
which he was more fluent.

Like Gerhard, Edith was a refugee from Germany; in fact, she was also
from Frankfurt. The families of both her parents had lived for many generations
in southern Germany. The family of Edith's mother, Alice Frohmann, was
rooted in Baden and in banking. Edith's paternal great-grandfather, Emanuel
Straus, and his father, Nathan, before him, living in the village of Tauber-
bischofsheim, had been providers to the court of Prince Franz Wilhelm von
Salm.[1] Her grandfather, Levi Straus, had entered into banking, as had others
in the family of Levi's wife, Rosalie. Levi and Rosalie's second son was
Edith's father, Friedrich, a urological surgeon, who married Alice Frohmann
and established a prominent practice in Frankfurt, where Edith's sister Lilian
was born in 1909 and Edith in 1910. After Friedrich died in 1915, his widow,
Alice, married Ernst Horkheimer, who worked in the bank of the Frohmann
family. Ernst was the father Edith knew during most of her youth in Frankfurt.

After Edith entered the Viktoriaschule to study art in 1929, economic
conditions dictated that she should earn some money; so she worked at her
grandfather's bank, Frohmann & Co., during the day and studied the history
of art at night, at the Johann Wolfgang von Goethe Universität in Frankfurt.
She was accepted as a volunteer at the bank Bondi and Maron in Dresden,
but just as she was ready to start, at the beginning of April 1933, the Nazi
government ordered a national boycott of Jewish stores and businesses. She
received a call from one of the head clerks telling her not to come to the

---

[1] Family background and incidents leading to her leaving Germany were relayed as personal communications
from Edith Schmidt.

bank but to meet him at the well-known Café Zunz instead. He told her they could not have her working at the bank. She did find work at a Jewish private bank, Baer and Auerbach, in Munich; but in 1935 Nazi thugs threw rocks that broke the bank's ground-floor windows; the owners closed the bank permanently.

A second incident in 1935 convinced her to flee Germany. She and a cousin were at lunch in a restaurant not far from the Braune Haus, the seat of the Nazi Party. "During lunch we overheard a heated discussion between one of the Party members in uniform and a waitress. "No Jews are permitted in this place," he screamed, 'Get them out!'" Deeply shaken by this experience, she packed two suitcases, which she sent home to Frankfurt, and a knapsack, which she took as she hitch-hiked from Munich, via Lake Constance and Switzerland, to Milan, Italy.

She took jobs as a governess in Milan, and then undertook studies of Italian literature, Italian history, and the history of art at the Universita per gli Stranieri in Perugia. It was there and then, in May 1937, that her path met that of Gerhard as he was visiting Perugia with his mother, in the two-month interval between his positions in Kingston and New York. They did not have a lot of time together, but when he left to spend some days in Naples, once again at the Marine Biology Institute, they agreed to stay in touch by mail. He returned to New York three weeks later. In the coming year he would carry out some of his most important experiments.

Figure 1. Dr. Morris Cynkin, Professor of Biochemistry at Tufts University School of Medicine, who conducted and recorded conversations with Gerhard Schmidt between 1971 and 1973.

Photo provided by Dr. Rosemary Polsky-(Cynkin)-Newman.

Figure 2. Participants of the 1973 symposium honoring Dr. Gerhard Schmidt on his election to the United States National Academy of Sciences: *Front row, left to right:* Fritz Lipmann, David Nachmansohn, Gerhard Schmidt, Erwin Chargaff, Henry Mautner. *Back row, left to right:* George Brawerman, Carl Cori, Severo Ochoa, Herman Kalckar.

Figure 3. Dr. Julius Schmidt, father of Gerhard Schmidt.

Figure 4. Isabella (Gombrich) Schmidt, mother of Gerhard Schmidt.

Figure 5. Dr. Julius Schmidt in a chemical laboratory at the Stuttgart Technical University.

Figure 6. Young Gerhard Schmidt with sisters Elizabeth and Marion.

Figure 7. Gerhard Schmidt at about age 23, with sisters Elizabeth, Marion, and Renate.

Figure 8. At Julius Schmidt's 60th birthday dinner, 1932. *Seated*: Elizabeth, Julius, Isabella, Marion. *Standing*: Gerhard and Sepp Grünbaum, husband of Elizabeth.

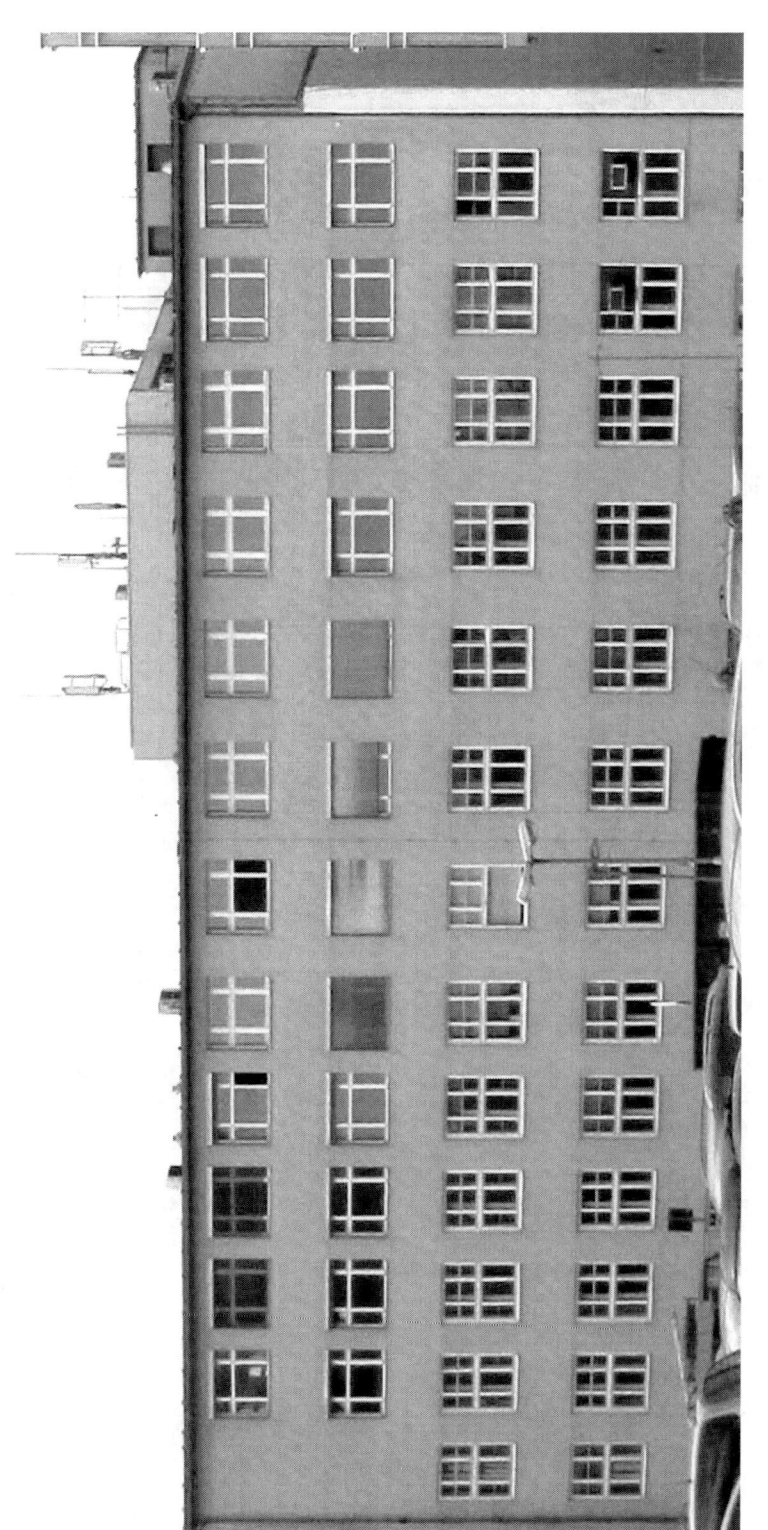

Figure 9. Gustav Embden Institute for Vegetative Physiology. A portion of the building remained in 2010 but was about to be removed and replaced with a new research building.

Figure 10. Professor Gustav Embden with students and staff. Dr. Embden is in the front row, fifth from the left (*arrow head*); Gerhard is in the back row, second from the left (without glasses).

Figure 11. Professor Bernhard Fischer-Wasels, chairman of the Department of Pathology, University of Frankfurt Medical School.

Figure 12. Edith Straus Horkheimer, who met Gerhard in Gubbio, Italy, in 1937.

Figure 13. Dr. Siegfried J. Thannhauser, professor and chairman of the Department of Medicine at Freiburg University in 1932, moved to the Boston Dispensary and Tufts University School of Medicine in Boston in 1935, and recruited Gerhard Schmidt to the Dispensary's Laboratory of Biochemistry.

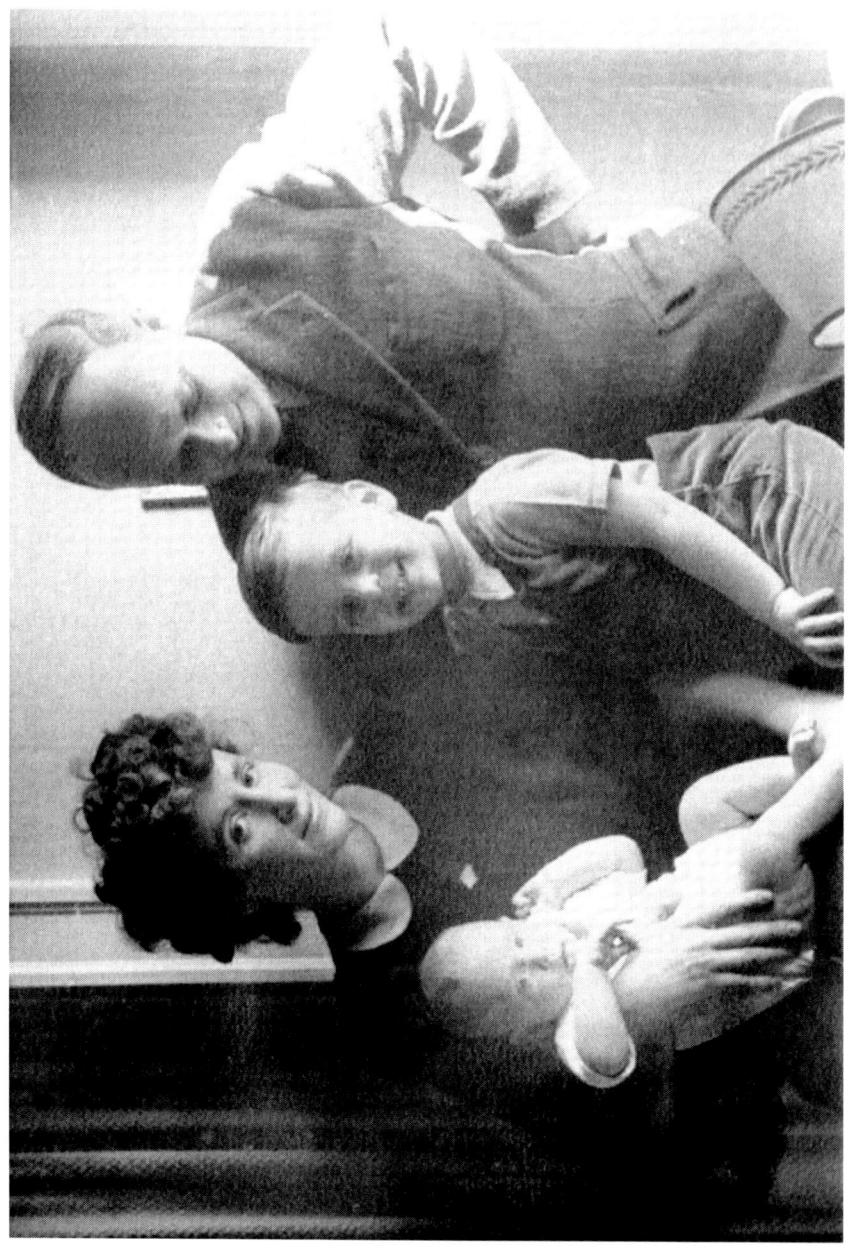

Figure 14. Gerhard and Edith Schmidt in Boston, with Michael and Milton.

Figure 15. Isabella Schmidt, Gerhard's mother, in Boston.

Figure 16. Gerhard Schmidt with horseshoe crab skeletons, collected for him by John Gulland at the 1947 Cold Spring Harbor Symposium.

Figure 17. Gerhard and Edith Schmidt with Michael and the family Studebaker.

Figure 18. Roy Keenan, Gerhard Schmidt, and Teruji Tanaka.

Figure 19. Maurice Bessman, Gerhard Schmidt, Marya Seraydarian, and Lowell Greenbaum; photo taken in the 1950s.

Figure 20. Gerhard with cello.

Figure 21. Maurice Bessman, Gerhard and Edith Schmidt, Lowell Greenbaum in the 1970s.

Figure 22. Herman Kalckar and Gerhard Schmidt, 1973.

Figure 23. Gerhard Schmidt, 1970s.

New York, P. A. Levene, and the Rockefeller Institute

W hen Gerhard's steamer docked in New York, he retrieved a bicycle, which he had bought in Italy and had brought with him on the boat. He rode the bicycle from the customs office on the Battery up Broadway to Times Square, where he bought a copy of *The New York Times* and scanned the apartment rental advertisements. He found a rooming house offering a room, with a shower in a shared bathroom, for five dollars a week (prices were that low at the time of the Great Depression). He followed that lead, saw the room and took it. Two days later he appeared at the Rockefeller Institute and amazed Dr. Levene with his report that he had found living quarters at 22 East 72nd Street, just off 5th Avenue, an address one would not expect for someone with a salary of $2000 per annum (even less than he had received in Kingston). In fact, Levene had been very worried about just where it might be when Gerhard told him first that he found a room for five dollars a week, warning that it is important to live in an appropriate neighborhood. Gerhard had simply looked for a place near the Institute.

Showing up for the July 1 beginning of his term, Gerhard found the Rockefeller Institute nearly empty of faculty and staff. He immediately obtained an appointment to see Dr. Levene, who told him that July and August were official vacation times, and he could very well start in September. Levene himself spent a lot of the summer in his country house in Connecticut, but would be in the department at least some of the time. Gerhard responded that he had just been on vacation in Italy, was eager to begin, and was especially interested in working on nucleic acids. Levene agreed he could in fact begin immediately even if there were not many other people on hand, and suggested that Gerhard study the enzyme *ribonuclease*, an RNA-degrading enzyme discovered 17 years earlier by Walter Jones.[1] The results of this project comprised one of Gerhard's important contributions to nucleic acid biochemistry.

---

[1] W. Jones, "The Action of Boiled Pancreas Extract on Yeast Nucleic Acid," *American Journal of Physiology* 52(1920): 203–07.

Ribonuclease is very unusual in that it recovers nearly all its catalytic activity after it is incubated in boiling water and then cooled. Virtually all enzymes known in 1937 were destroyed irreversibly by such heating. In fact, this behavior made some biochemists skeptical as to whether the RNA-degrading activity really was a protein enzyme. Levene had been among the skeptics and, years earlier, he had also been critical of Walter Jones's theory of nucleic acid structure. In 1937, however, Levene's Rockefeller Institute colleague, René Dubos, confirmed Jones's report of the remarkably heat-stable enzyme activity,[2] a finding that stimulated Levene to study it further, perhaps in part so that he would have a chance to commend Jones's work after all; Jones had died in 1935. In fact, the paper that Gerhard and Levene published the following year was entitled *"Ribonucleodepolymerase:* the Jones–Dubos Enzyme."[3] More important as a motivation, however, Levene was hopeful that the enzyme would provide a way to isolate each of the four nucleotides of RNA without requiring harsh chemical conditions. As noted earlier, the alkali degradation that was in general use until then yielded an adenylic acid that differed from the adenylic acid isolated from muscle tissue by having its phosphate ester at a different position on the ribose. Perhaps enzymatic digestion would yield the more natural isomer.

Although Gerhard and Levene did confirm the existence of the enzyme, their results differed in an important aspect from those of Jones, who had reported that the product of enzyme degradation was a solution of mononucleotides. Gerhard found that the products of maximum digestion by the enzyme were still polymerized; the enzyme stopped far short of producing mononucleotides. In fact, about 60% of the products were large enough to be precipitated by hydrochloric acid. Even most of the digestion products that were acid-soluble were larger than mononucleotides, as they could not pass through a dialysis membrane, whereas mononucleotides could do so. Thus Levene's hope for an enzymatic production of mononucleotides was not realized.

This first demonstration of limited digestion by the RNA-degrading enzyme added to the growing concept that nucleic acids are large polymers, but the conclusion was still framed in terms of the tetranucleotide hypothesis: "The function of the enzyme is that of a depolymerizing agent, limited to the dissociation of the tetranucleotides of high molecular weight into those of lower molecular weight. Native ribonucleic acid is a polymer of the tetranucleotide." Later, in his own laboratory at Tufts University, Gerhard determined the sites at which *ribonuclease* cleaves the nucleic acid chain, helping to explain why it does not break the RNA down to single nucleotides.

---

[2] R. J. Dubos, "The Decomposition of Yeast Nucleic Acid by a Heat Resistant Enzyme," *Science* 85(1937): 549.
[3] G. Schmidt and P. A. Levene, "Ribonucleodepolymerase (The Jones-Dubos Enzyme)," *Journal of Biological Chemistry* 126(1938): 423–34.

Gerhard wrote a manuscript describing these experiments and Levene sent it to the *Journal of Biological Chemistry*. A short while later, Levene appeared in the laboratory, rather upset, wringing his hands, saying "This is the first time the editor of the *Journal of Biological Chemistry* has returned a manuscript to me for modification!" Levene's writing was very precise. He was used to having his articles accepted without changes.

It is interesting to look back and see the large scale and methods with which these experiments were performed. To follow the rate of enzyme activity by measuring the decrease that did occur in the amount of acid-insoluble material, Gerhard used solutions of 100 mg/mL of yeast RNA and many mL of enzyme solution. To obtain the acid-soluble and acid-insoluble fractions of the products, he carried out the digestion of 20 g of RNA with 1000 mL of enzyme solution! He still applied Embden's gravimetric method for the measurement of total phosphorus content of combusted RNA or digestion products as well as the methods he had developed in Frankfurt for purine determination. Emerging technology would allow analysis of much smaller amounts of material. Dialysis, important in this project, was not yet widely applied; the semipermeable membrane used by Gerhard for this purpose was a kind of paper prepared by the company Schleicher and Schuell.

AMONG BRIGHT STARS AT THE ROCKEFELLER INSTITUTE

Gerhard found a very stimulating atmosphere at the Rockefeller Institute, generated by the outstanding scientists who led various departments, the colleagues and trainees they attracted, and the numerous seminars by visiting scientists. He was impressed by the lineup that included microbiologist René Dubos; peptide chemist Max Bergmann and his associates William Stein, Stanford Moore, and Klaus Hofmann; physiologist Alexis Carrel; chemist and enzymologist Leonor Michaelis; clinical chemist Donald Van Slyke ("the most pleasant and most humble of them all"); cell biologist/biochemist Albert Claude; immunochemist Karl Landsteiner; physical chemist Alexander Rothen; and the Director of the Institute, Herbert Spencer Gasser, who was awarded the Nobel Prize for discovery and application of the cathode oscillograph in physiology. Indeed, many of the Rockefeller Institute scientists were or went on to be Nobel Prize winners and/or members of the National Academy of Sciences.

There were cultural behaviors to be learned in this assembly of stars. Not long into his term at the Institute, Gerhard heard a stimulating seminar by

Max Bergmann,[4] a peptide chemist, who addressed the questions of both chemical peptide synthesis and peptide hydrolysis. In the discussion, Leonor Michaelis[5] posited the suggestion that biosynthesis may result from reversibility of the action of hydrolytic break-down enzymes, such as *trypsin*, which, as Bergmann showed, could catalyze the formation of some amino acid anilides. Feeling very free in this American environment, Gerhard rose to state that there are many examples in biological systems in which biosynthesis is not the reverse of hydrolysis but, rather, follows a very different pathway. The next morning he found a message in his laboratory, asking him to go to see Joseph Fruton,[6] then a junior scientist at the Institute. Fruton, hoping Gerhard would accept a well-meant suggestion, advised him that it did not look too good for a newcomer to the Rockefeller Institute, of a fairly young age, to contradict some outstanding expert in physical chemistry such as Michaelis, and it might be better for one's future to be a bit more cautious in joining in these seminar discussions.

The hierarchical structure at the Institute was evident also at lunch, which was served in a large magnificent room that was set up with the hope that people from different labs and departments would mingle and exchange ideas; it was recognized that there was not a lot of such contact between departments. In spite of the hope, one did not sit in any old place of one's choosing. Usually the members of a department came in together, in a way that reminded Gerhard of a scene from Wagner's Tannhäuser and the Singers' Contest at

---

[4] Max Bergmann (1886–1944), born in Fürth, Germany; studied for thr PhD with Emil Fischer, 1911, was director of Kaiser Wilhelm Institute for Leather Research, Dresden, 1921–33; emigrated to the United States to the Rockefeller Institute. His research was on carbohydrates, amino acids, chemical synthesis of peptides and the specificity of proteases.

[5] Leonor Michaelis (1875–1947), born in Berlin, earned an MD (Freiburg) and studied for his doctorate (Berlin) with Paul Ehrlich with a thesis on immunological precipitins. He carried out varied studies on the Wasserman test reagent, electrophoresis, theory of electrolyte dissociation, and colored pH indicators, but is best known for his theory of enzyme activity (with Maud Menten) and enzyme inhibitors. He moved to Nagoya in 1922, to Johns Hopkins 1926–29, and was at the Rockefeller Institute from 1929 on, working on oxidation/reduction, heavy metal catalysis of free radicals; L. Michaelis, "An Autobiography with Additions by D. A. MacInnes and S. Granick," *National Academy of Sciences Biographical Memoirs* (Washington, DC: National Academy of Sciences, 1958). Available online at http://www.nasonline.org/publications/biographical-memoirs/memoir-pdfs/michaelis-leonor.pdf. In 1921 Michaelis reported that some work of Abderhalden could not be repeated. It has been stated that his move to Nagoya occurred because his academic career in Germany was compromised after this report; Ute Deichman and Benno Müller-Hill, "The Fraud of Abderhalden's Enzymes,"*Nature*, 393(1998): 109–11.

[6] Joseph Fruton (1912–2007), born in Poland, lived in the United States, then Russia, and returned to the United States. He earned a BS in 1931 summa (Columbia) and a PhD from Columbia College of Physicians and Surgeons in 1934; worked at the Rockefeller Institute with Max Bergmann 1934 to 1945, and then joined the faculty at Yale. He was known for work on proteolytic enzymes and chemical synthesis of peptides, as well as a textbook written together with his wife, Sophia Simmonds. He also became a historian, publishing a notable book in 1990: *Contrasts in Scientific Style: Research Groups in the Chemical and Biochemical Sciences*; M. Singer, "Joseph Fruton: A Biographical Memoir," *National Academy of Sciences Biographical Memoirs* (Washington, DC: National Academy of Sciences, 2009). Available online at http://www.nasonline.org/publications/biographical-memoirs/memoir-pdfs/fruton-edwin.pdf.

Wartburg, in which "every choir comes in with its coat of arms and with its flag bearer and its followers and they all converge through the various doors into a tremendous hall where the competition takes place." Each group came to lunch following the department chief. There was little or no intermingling until, perhaps, some exchanges occurred after lunch.

A special effort was needed to enhance communication among geographically separated parts of the Rockefeller Institute, which included a division on the campus of Princeton University. At an annual visit of New York scientists to the Princeton Division, Gerhard was introduced to laboratories of the enzymologist John H. Northrop[7] and structural biologists Wendell Stanley[8] and Hubert Loring, additional examples of scientists breaking new frontiers; but apart from the annual visit, there did not seem to be a lot of exchange between the New York and Princeton scientists. Gerhard and Morris Cynkin, his interviewer, explored together the reasons for this lack of communication and its main consequences, that is, what they perceived as a relatively slow diffusion of knowledge and new methods. Although New York had an abundance of superb labs in many institutions, Gerhard was struck by the difficulty in getting from one place to another to visit and hear a seminar, especially in the time crunch that was a new experience for him but was the common pattern for his new colleagues in America. He saw that many faculty members had long commutes to work and wanted to be with their families in the evening, so they wanted to use each day efficiently; going to hear a talk at another institution would take up a good part of the day.

FINDINGS THAT OVERTHROW A MENTOR'S HYPOTHESIS

Gerhard found many stimulating associations at the Institute, and one of them led to another most important piece of work. A visiting-scientist seminar was presented by Einar Hammarsten, Professor of Chemistry at the Karolinska Institute in Stockholm, whom Gerhard had met during his brief stay in Sweden. Hammarsten had brought with him a sample of DNA prepared by a method he had described in 1924.[9] For the intervening 13 years, biochemists had not taken notice of that method even though it yielded more pure and less degraded DNA than anyone else had obtained—another example of slow

---

[7] Northrop purified and crystallized the enzyme *pepsin* in 1929, proving its protein nature; he shared the 1946 Nobel Prize in Chemistry.

[8] Stanley worked with Northrop on *pepsin* crystallization, sharing the 1946 Nobel Prize, and later they worked on crystallized tobacco mosaic virus.

[9] E. Hammarstein, "Zur Kenntnis der biologischen Bedeutung der Nucleinsaureverbindungen *Biochemische Zeitschrift* 144(1924): 383–469.

diffusion of knowledge and methods, in this case probably reflecting the tendency of scientists to be invested in their own concepts and methods and slow to accept new ones. Hammarsten's preparation was an improvement over that based on an earlier finding by Bang,[10] who showed that exposure of nucleohistone to alkali separates the protein and nucleic acid components from each other, allowing the separate precipitation of the DNA without any breakdown; unlike RNA, DNA is resistant to alkali. Hammarsten separated the DNA and protein with a solution of high ionic strength, at neutral pH and low temperature, and then precipitated the DNA. When dried, the DNA was fibrous. When it was dissolved in water or dilute saline, it formed a very viscous gel, even at low concentration, a reflection of its high degree of polymerization—that is, its high molecular weight. The gel was colorless and clear. It was a highly purified sample of DNA that had not been exposed to depolymerizing chemicals.

Levene obtained some of the DNA from Hammarsten. Gerhard was excited by seeing DNA of such a degree of polymerization and sought permission from Levene to use some of it in experiments similar to those he performed in Kingston. He would determine how much phosphate in that DNA could be released by *monophosphatase*, which can cleave only the phosphate ester at the end of a chain. He tested four samples: the preparation by Hammarsten, which had been reported to have a molecular weight of about 1 million; a sample obtained using an earlier preparative method of Neumann; one prepared by Levene; and one prepared with enzymatic depolymerization using Feulgen's method. Purified intestinal *alkaline phosphatase*, free of depolymerizing activity, released no detectable phosphate from the Hammarsten preparation, even with 4 hours of incubation. That meant that the polymer was so long that the one terminal phosphate was an unmeasurably small fraction of all the phosphate in the chain of nucleotides. With the Neumann preparation, 8% of the phosphate was released by the *phosphatase*; with the Levene preparation about 20% was released, and with the Feulgen preparation, 47%, almost equal to a control 53% for a standard monoester, glycerol-phosphate. That is, the amount of phosphate ester accessible to the *phosphatase* enzyme was totally dependent on the molecular weight, the degree of polymerization of the polynucleotide. The lack of accessibility in the Hammarsten preparation meant that this DNA could not be a tetranucleotide, in which 25% of the phosphate would be at the end, or even a simple aggregate of tetranucleotides.

A simultaneous but different personal association added an important new dimension to this study. The Rockefeller Institute for Medical Research was also home for The International Health Division of the Rockefeller Foundation,

---

[10] O. Hammarsten and S. G. Heden, *A Textbook of Physiological Chemistry*, 7th ed., trans. J. A. Mandel (New York: John Wiley, 1914), 367. Retrieved online at http://books.google.com/books.

in which experiments were being carried out by Edward G. Pickels with a new kind of ultracentrifuge.[11] Pickels was willing to test the behavior of nucleic acid samples in this centrifuge, in which a photographic record shows the progressive movement of a light-absorbing or light-diffracting material to the outer part of the small chamber in which the solution is placed when the chamber is subjected to high centrifugal force. The results confirmed that the Hammarsten material was of uniformly high molecular weight. The Neumann material sedimented with a blurred boundary, reflecting the presence of heterogeneous mixture of molecules of different size, ranging from about 50,000 to over 1 million. The Levene and Feulgen preparations were too small to be sedimented at 54,000 rpm (210,000 x g) for 35 minutes. These results confirmed the relationship between molecular size and accessibility of chain-terminal phosphate esters. The prevailing conclusions on DNA structure, in particular the tetranucleotide hypothesis, had been developed based on experiments with DNA that had been severely degraded during its isolation. The Hammarsten DNA had not been seriously degraded.

These data, obtained in Levene's own laboratory, spelled the end of his tetranucleotide hypothesis. Gerhard recalled that Levene accepted that outcome, though he still held on to the idea of some regularity in nucleotide composition of the chain. In his earlier book on nucleic acids, published in 1931,[12] Levine had included a possibility that DNA could be a polymer of tetranucleotides. Gerhard and Levene published a preliminary note on these results in the journal *Science* and then a full paper in the *Journal of Biological Chemistry*, both appearing in 1938.[13]

YET MORE MOVES WILL BE REQUIRED

Gerhard was thoroughly enjoying his year in New York, impressed both by the caliber of science at the Institute and by friendships of people within and beyond the laboratories. He re-established contact with the harpsichordist Yella Pessl Sobotka, to whom Levene had introduced him two years earlier, and through Levene and Sobotka, met others with whom he could play chamber music. Within Levene's department, Gerhard particularly enjoyed the company of analytical chemist Adelbert Elek, from Hungary, and biochem-

---

[11] J. H. Bauer and E. G. Pickels, "An Improved Air-Driven Type of Ultracentrifuge for Molecular Sedimentation," *Journal of Experimental Medicine* 65(1937): 565–86.

[12] P. A. Levene and L. W. Bass, *Nucleic Acids* (New York: Chemical Catalog Co., 1931).

[13] G. Schmidt and P. A. Levene, "The Effect of Nucleophosphatase on "Native" and Depolymerized Thymonucleic Acid," *Science* 88(1938): 172–73; G. Schmidt, E. G. Pickels, and P. A. Levene, "Enzymatic Dephosphorylation of Deoxyribonucleic Acids of Various Degrees of Polymerization," *Journal of Biological Chemistry* 127(1939): 251–60.

ist Gene M. Meyer,[14] a witty, wealthy, and very kind colleague, who would soon play an important role in Gerhard's life by offering to be guarantor for his immigration visa application. Gerhard had achieved important discoveries in just one year at the Rockefeller Institute and would have been pleased to continue his work there. He had been offered his position for two years, but he and other staff members were notified by the administration of the Institute that Levene's department would be closed at the end of 1938. Levene was then 69 years old. By the end of the year Gerhard would have to identify his next move.

Meanwhile, the situation in Europe was causing him more and more anxiety. In March 1938, Hitler was welcomed into Vienna as Nazi Germany took over Austria in the "Anschluss"—the integration of German peoples. Austrians began more vigorous assaults on Jews and Jewish property than had been seen even in Germany. In Italy, Mussolini and the Fascist regime also made a turn toward increasingly anti-Jewish policies. The situation was becoming worse both for Gerhard's mother and sisters and for Edith Straus, whom he had met in Perugia and with whom he had continued to exchange letters.

After completing her studies in Perugia but before her final exams, Edith moved to Naples to take a job in a major bookstore, Detken and Rocholl.[15] Between May 5–10, 1938, Hitler visited Italy, and Naples was on the itinerary. The owner of the bookstore where Edith worked, Bernhard Johannowsky, a Sudeten German and Nazi sympathizer, planned to give Hitler a gift of a set of lithographs, and asked Edith to prepare them in a folder decorated with the Italian flag and the swastika intertwined. Unsettled, she completed that task. Shortly after she returned home, two police officers came to her door, and took her into "protective custody," a precaution the regime was taking before Hitler's visit by rounding up aliens, particularly German refugees, who might cause some kind of disturbance. Though one of the officers treated her kindly, they took her to the State Prison of Poggio-Reale, where she was held for an agonizing eleven days, together with about twelve German Jewish women and a small group of German prostitutes. On one of those days she was taken for an interrogation by two police officers, in a small room in which she was separated from the interrogators by an iron grill. They asked her why she was in Italy and examined all her activities. The questioning officer was the one who had treated her well during the arrest. When his companion left, he explained that this process was sort of a ruse. Being worried about her,

---

[14] G. M. Meyer published several articles on glycosides with P. A. Levene in *Journal of Biological Chemistry* between 1928 and 1938.

[15] The following information on Edith is from E. Schmidt, *Some Episodes from My Life*, prepared for Steven Spielberg's documentation of Holocaust survivors. Her testimony, with more information about her family and personal history, is recorded in the University of Southern California Shoah Foundation Archives; available at http://vhaonline.usc.edu/viewingPage.aspx?testimonyID=33501&returnIndex=0#.

he had brought some chocolates that some nuns would transmit to her. He also offered his phone number in case she would need help. When discharged, she was required to report daily to the police station.

She went to the main post office to place a call to her parents in Frankfurt, feeling that either sending a letter or calling from a private phone was too risky. When her mother answered Edith said only one word: "Come." Two days later her parents arrived in Naples and they sailed to Capri, where they could talk freely on the beach. Her stepfather, Ernst Horkheimer, had taken steps so that he and Edith's mother, Alice, became British citizens. Together, they all traveled by train to Florence, where the British vice consul promised to provide Edith a visa as soon as she finished taking her exams. A few months later she did take the final exams in Perugia and in September 1938 she left Italy for England. On the way she had a visa granted by Switzerland, limited to a one-day stopover, so that she could visit her uncle Moritz (her father's brother) and aunt Lilly, then in Basel. Crossing France, on the other hand, she was confined to the train as she could not obtain a visa even for a stopover in that country. She crossed from Calais to Dover where she was welcomed by her parents, who had come from Frankfurt. She finally felt secure. Her parents then returned to Frankfurt; but, two months later, after Kristallnacht, with the Germans pursuing him because he had given shelter to a Jewish man for whom the Gestapo was searching, Ernst Horkheimer flew from Frankfurt, never to return. Edith's mother, Alice, left soon afterward. Her sister Lillian and Lillian's husband, John, who had just begun a position as rabbi of a Heidelberg synagogue, came a few months later, after making certain that the children of their congregants had reached safety in England. The family, including her brother Mickey (Milton) lived in Golders Green, in northern London. Mickey, 18 years old at that time, became a pilot with the Royal Air Force. At the age of 23, he was killed when his plane was shot down over Germany.

In London Edith refreshed her knowledge of English with a Berlitz school course and began to work as a private secretary to Anna Schwab, who both opened her home to refugees and supported the German Jewish Refugees Committee. The Committee worked with the British Home Office to get prominent endangered German and Austrian Jews to England. In the next year, after the outbreak of war, Edith was promoted to a position at the housing bureau, where she was involved in placing refugee children in homes and hostels and eventually arranging for their evacuation from London to the countryside.

As a further consequence of the Nazi takeover of Austria in the Anschluss of March 1938, increasing numbers of Jews were seeking to leave Germany and Austria, forming long lines at U. S. embassy and consular offices. Gerhard became concerned with his own status, as he was holding just a student visa in the United States. A lawyer friend warned him that, with the expiration of his visa, he could be deported at the end of his term at the Rockefeller Institute. Regretting that he had not applied for an immigrant visa when he was moving from Kingston to New York, he quickly began to assemble what was required for a permanent visa. He needed an affidavit from someone guaranteeing that he would have a source of money and not become dependent on public funds. Dr. Gene Meyer, in Levene's lab, happily provided the affidavit, jokingly saying, when he agreed, "But all I'll be responsible for is two dollars a week." The next requirement was to obtain police certificates from all the countries in which he had resided—Germany, Italy, Sweden, and Canada—showing no history of arrests. He worried, particularly, whether he could obtain one from Germany. With the help of the Rockefeller Institute, he was in fact able to get all of them. Then he realized that his German passport was near its expiration. He had to go himself to the German Consulate to seek an extension. He approached the Consulate, located in the Battery, just as the tail of the great hurricane of September 21, 1938, was passing through New York. He had to wait for an hour till it was possible to cross the street from the subway station to the building he had to reach; he got there just before closing time. Although it was the autumn of 1938, the consular office did extend his German passport.

There was still another issue. He could not obtain his U. S. visa while he was in the United States. Immigration visas are issued only by consulates of the United States, not by any office within the country. He thought he could go back to Canada, but was informed by Canadian officials that he could not enter that country without having a valid permanent visa for return to the United States. The nearest country that did not require such a return document was Cuba. So he traveled by bus to Miami, a 31-hour ride, and there boarded a boat for Cuba, where he spent eight days in the preparation and processing of his visa application. It was easy to have the German and Italian certificates translated into English, but somewhat more difficult to find someone to translate the Swedish document. He finally submitted all the required documents. Apart from waiting for bureaucracy, these were not difficult days. He had time for marvelous bus rides to the countryside and for leisurely breakfasts at outdoor cafes, benefitting from having carefully saved money from his stipends and from low costs in Cuba. After a prolonged breakfast he would smoke a cigar, feeling "I only wish the Nazis could see me now—a German Jew, forced out of Germany—smoking two Havana cigars" (very expensive in Germany but only a few cents in Cuba). He obtained his U.S. immigration visa and could return to New York.

Twelve

St. Louis with the Coris

In the spring of 1938, when informed of the pending closure of Levene's department at the end of the year, Gerhard explored the possibilities for his next move. He knew nothing definitive could be arranged before he obtained his immigration visa, but he wrote letters. One was to Dr. Siegfried Thannhauser, Professor of Medicine at Tufts University School of Medicine in Boston.[1] Dr. Thannhauser had been Professor of Medicine in Freiburg, Germany, an established luminary in both clinical medicine and biochemistry; but, being Jewish, he was dismissed from his post by the Nazi regime in 1934. He arrived at the Pratt Diagnostic Clinic of the Boston Dispensary (which had become part of the New England Medical Center) and Tufts in 1935, and there established a biochemistry laboratory. Gerhard had been introduced to him while both were still in Germany, where Thannhauser had made important discoveries in nucleotide chemistry. Gerhard also wrote to Dr. Carl Cori at Washington University School of Medicine in St. Louis, knowing that both Carl and Gerty Cori were interested in carbohydrate biochemistry, the field in which Gerhard's mentor, Gustav Embden, had made such important contributions.

Thannhauser sent an encouraging reply and invited Gerhard to visit Boston. With help from the Rockefeller Institute, which was supporting the job searches of people in Levene's lab, Gerhard travelled to Boston during the summer, an overnight trip by coastal steamer, through the Cape Cod Canal; it was the same kind of trip he had made in the other direction, to New York, after attending the International Congress of Physiology in Boston in 1929. Thannhauser offered him a research associate position, with a stipend of $2,500; but Gerhard's understanding was that if he had accepted, it would have required dismissal of a technician, Joe Penotti, then working with Thannhauser. That was an unattractive idea, especially at a time when it would have been difficult for Penotti to find a new job.

At about the same time, Gerhard received a positive response from Carl Cori as well, saying it was very likely that a position could be arranged in St. Louis, as long as Gerhard was still interested in carbohydrate biochemistry.

---

[1] Information on Dr. Thannhauser's life and work is presented in Chapter 13.

It was not, in fact, a field in which Gerhard had worked since his early years with Embden, but he had been impressed by the recently published discovery, by the Coris, of glucose-1-phosphate, a product and precursor in glycogen metabolism,[2] and he certainly had been interested in the varying sensitivity of different kinds of phosphate esters to hydrolysis by acid, base, or enzyme catalysis. There was, therefore, a scientific basis for accepting the offer from St. Louis, without the prospect of displacing another person. Late in the autumn of 1938 he moved to St. Louis.

Though he spent just over a year in St. Louis, he achieved and published significant scientific results and established long-lasting friendships. Like Gerhard, Carl and Gerty (Radnitz) Cori had European origins. Both were born in Prague in 1896. Both earned their medical degrees where they met, at the medical college of the German University of Prague in 1920. They married and, after short stays in Vienna and Graz, moved to the United States in 1922, where Carl became biochemist at the State Institute for Studies of Malignant Disease—later renamed the Roswell Park Memorial Institute—in Buffalo, New York, and arranged a position for Gerty as assistant in pathology. They moved to St. Louis in 1931, when Carl became professor of pharmacology at Washington University, where he later served as professor and chairman of the Department of Biochemistry and where Gerty became a research associate. Carl and Gerty worked together in science, first on immunological studies of "complement" (a complex set of proteins in blood plasma that, together with specific antibodies, lead to destruction of foreign cells such as bacteria or blood cells and to inflammation) when they were still in Austria and, in the United States, on the effects of insulin and epinephrine on glucose metabolism, and then on the biochemistry of glycogen breakdown and formation. For their work on glycogen and glucose metabolism they would share the 1947 Nobel Prize in Physiology or Medicine, along with Bernardo Houssay,[3] who had discovered the effect of anterior pituitary hormone on sugar metabolism (and Gerty was named professor of biochemistry in that year). In Gerhard's recollection, Gerty was especially intellectually stimulating. She was also "very frank, very warm-hearted, and had great understanding for the situation of a refugee from Germany."

---

[2] C. F. Cori, S. P. Colowick, and G. T. Cori, "The Isolation and Synthesis of Glucose-1-Phosphoric Acid," *Journal of Biological Chemistry* 121(1937): 465.

[3] Bernardo Houssay, a native of Buenos Aires, became professor of physiology in the medical school at Buenos Aires University in 1919. He also organized the Institute of Physiology at the medical school, and remained professor and director of the Institute until 1943, when he was removed from his post because of political views favoring democracy. He then founded and directed the Instituto de Biología y Medicina Experimental. From "Bernardo Houssay—Biography," *Nobel Lectures, Physiology or Medicine 1942–1962* (Amsterdam:Elsevier, 1964). Retrieved November 2, 2012, from Nobelprize.org. http://www.nobelprize.org/nobel_prizes/medicine/laureates/1947/houssay-bio.html.

GLYCOGEN PHOSPHORYLASE

The scientific finding that attracted Gerhard to the Cori lab had just been published in 1937, by the Coris together with Sidney Colowick. Colowick was a St. Louis native who had just joined the Cori lab after earning a Bachelor of Engineering degree in 1936. He became a graduate student of the Coris, receive his PhD in 1942, and went on to an illustrious career in biochemistry, much of it at Vanderbilt University.[4]

As described in the 1937 publication, the Coris and Colowick discovered that glycogen breakdown begins with cleavage of glucose residues from linear glucose chains by the process of phosphorolysis. This process, given its name earlier by Jakob Parnas,[5] is the addition of phosphoric acid across the broken phosphate ester bond. Phosphorolysis of glycogen, yielding glucose-1-phosphate as a product, is catalyzed by the enzyme *glycogen phosphorylase*, which they obtained in only a partially purified form. Their enzyme preparation contained an important second enzyme, later named *phosphoglucomutase*, which converted the glucose-1-phosphate to glucose-6-phosphate. The ability to supply free glucose to the blood from glycogen depends on these two enzymes and a third one, *glucose-6-phosphatase*, which catalyzes a hydrolytic cleavage of the phosphate from the glucose—that is, by addition of water across the broken bond. The identification of steps from liver glycogen to blood glucose was a landmark discovery in the regulation of sugar metabolism.

When Gerhard arrived in the lab, his task was to purify the muscle *phosphorylase* so that it was free of the *phosphoglucomutase* activity, allowing its mechanism to be studied and eventually facilitating crystallization of the enzyme protein. Soon after his arrival, he achieved the separation by absorption of the mixture onto insoluble aluminum hydroxide, followed by elution of the *phosphorylase* alone by di-sodium phosphate. With this purified enzyme, he made the important discovery that the enzyme-catalyzed reaction was reversible; that is, glucose-1-phosphate could be converted to a linear polymer of glucose, with release of inorganic sodium phosphate. Adenylic acid was required for both glycogen breakdown and poly-glucose synthesis—and could not be replaced by other nucleotides. Nor could glucose-1-phosphate be replaced by other sugar-phosphate esters. Although the cellular mechanism for natural glycogen synthesis from glucose-1-phosphate follows a more complex

---

[4]Biographical information on Sidney Colowick and discussion of his determination of the structure of the coenzyme NADH are presented in N. Kresge, R. D. Simoni, and R. L. Hill, "The Structure of NADH: The Work of Sidney P. Colowick," *Journal of Biological Chemistry* 280(2005): e36–e37.

[5]J. K Parnas and T. Baranowski, "Sur les phosphorylations initiales du glycogene," *Comptes Rendus des Séances de la Société de Biologie* 120(1935): 307–10.

pathway and yields a more complex polymer,[6] this reaction was the first enzymatic synthesis of a polysaccharide. Gerhard's initial results were presented at a meeting of the Missouri branch of the Society for Experimental Biology and Medicine on April 12, 1939, and fuller descriptions appeared in articles in the May issue of the journal *Science*, and an August issue of the *Journal of Biological Chemistry*.[7]

Following up the published work, Gerhard began some experiments on the kinetics of the *phosphorylase* reaction, observing a first-order kinetic reaction, but he did not have time to pursue those studies in depth. He also attempted to purify the *phosphatase* that releases free glucose into the blood from *glucose-6-phosphatase*; but time ran out for his stay in St. Louis. He had received an offer of a position without a time limit, from Siegfried Thannhauser in Boston, and he accepted it.

ORIGINS OF LONG-LASTING FRIENDSHIPS

While still in St. Louis he met Herman Kalckar,[8] then a fresh doctoral graduate from Copenhagen who was visiting the Coris while on his way to a postdoctoral research position at California Institute of Technology. The Coris had invited Kalckar to visit because they had seen two publications that resulted from his doctoral thesis work, carried out in the laboratory of Einar Ludsgaard in Copenhagen and published in the journal *Enzymologia* in 1937. Kalckar had demonstrated a greatly increased rate of phosphorylation—conversion of inorganic phosphate into phosphate bound to an organic compound—when cell extracts were incubated with aeration, providing plenty of oxygen. The phosphorylation was accelerated when certain substrates, such as malate,

---

[6] L. F. Leloir and C. E. Cardini, "Biosynthesis of Glycogen from Uridine Diphosphate Glucose," *Journal of the American Chemical Society* 79(1957): 6340–41; L. F Leloir and S. H. Goldemberg, "Synthesis of Glycogen from Uridine Diphosphate Glucose in Liver," *Journal of Biological Chemistry* 235(1960): 919–23.

[7] C. F. Cori, G. Schmidt, and G. T. Cori, "The Synthesis of a Polysaccharide from Glucose-1-Phosphate in Muscle Extract," *Science* 89(1939): 464–65; G. T. Cori, C. F. Cori, and G. Schmidt, "The Role of Glucose-1-Phosphate in the Formation of Blood Sugar and Synthesis of Gycogen in the Liver," *Journal of Biological Chemistry* 129(1939): 629–39.

[8] Herman Moritz Kalckar (1908–91), was born in Copenhagen into a Jewish Danish family, earned his MD in Copenhagen in 1933, and PhD with Einar Lundsgaard in 1939, associated with Fritz Lipmann on cell-free oxidative phosphorylation. In 1939, he received a Rockefeller fellowship for training at the California Institute of Technology, visited the Cori lab on his way, and returned in 1940 as research fellow. In 1946 he returned to Copenhagen, performed research on nucleoside and nucleotide metabolism, and mentored Morris Friedkin and Walter McNutt (later both professors at Tufts University) as research fellows. He studied the role of sugar nucleotides in galactose metabolism, and provided a basis for understanding galactosemia. In 1958 he moved to Johns Hopkins and in 1961 to Harvard Medical School. In 1979, he became distinguished professor of chemistry at Boston University. He wrote the biographical memoir on fellow National Academy of Sciences member Gerhard Schmidt; Eugene P. Kennedy, "Herman Moritz Kalckar 1908–1991," *National Academy of Sciences Biographic Memoirs*, available at ttp://www.nap.edu/readingroom.php?book=biomems&page=hkalckar.html/.

were added to the extracts. This was the first demonstration of what later became known as oxidative phosphorylation, the major process by which the potential energy present in food is captured by formation of the standard metabolic currency of energy, ATP. Carl and Gerty Cori were greatly impressed by this finding, which may have been what Gerty Cori had called a "missing link." In their groundbreaking work on both the breakdown and synthesis of glycogen and its role in regulation of blood glucose levels, the Coris recognized the importance of the addition of phosphate to glucose in the direction of synthesis. By then Meyerhof had shown there was an enzyme, *hexokinase*, which could transfer a phosphate from ATP to glucose to form glucose-6-phosphate, but Gerty Cori said *hexokinase* still could not account for all of the measurable additions of phosphate to organic ester form; there was still something missing, and this newly described oxygen-dependent process might be the answer. She was "tremendously excited, but also tremendously skeptical. She showed everyone this paper, and her final word was: 'I don't believe it.'"

Kalckar visited the Cori lab in February 1939. The secretary of the Department of Pharmacology in St. Louis arranged that Gerhard would meet Kalckar at the corner of Delmar Boulevard and Euclid Avenue. Gerhard did not know how he was going to recognize this visitor, and he was a bit late for the appointed time. Kalckar, however, was perceptive enough to guess that the rushing Gerhard was the person he was to meet. Gerhard saw "this dapper and handsome Danish aristocrat approach me and ask something like, 'Dr. Schmidt, I presume?'" That meeting was the beginning of a lifelong friendship.

Kalckar presented details of his work and convinced the Coris that it was flawless. Gerhard said, "I was thoroughly impressed with this work, because I encountered this issue not only at the Coris (thump on table for emphasis); but this question hung like a cloud over Embden's work during the 10 years I was there, because Embden and Laquer—his first associate professor—knew that muscle forms lactic acid from glycogen and it forms it from fructose diphosphate, but doesn't form it from glucose—and why? And this of course then gave rise—I think I mentioned that—to the hypothesis of the so called 'active glucose'—some form of gamma glucose—as the active form of the blood sugar—as the metabolically active form. Anyway, I mean, this problem was really solved for the first time in these two papers by Kalckar; he did nothing else but grind these kidneys up with a little sand and filtered it to get an extract—and it was really one of the most dramatic experiences in biochemistry I ever had, particularly as some of the active participants were such immortal persons like Gerty Cori."

In 1940 Kalckar would return to the Cori lab as a research fellow—as Gerhard's successor in fact—and collaborate with Sidney Colowick in the

discovery of an enzyme, *adenylate kinase*, which transfers a phosphate from
ATP to adenylic acid, yielding two molecules of adenosine diphosphate (ADP),
the substrate for oxidative phosphorylation. He and Gerhard remained connected throughout Kalckar's career in New York, Copenhagen, Baltimore,
and eventually in Boston in 1961, where he succeeded Fritz Lipmann as
director of the Biochemical Research Laboratory of the Massachusetts General Hospital.

In St. Louis Gerhard spent a lot of time with Sidney Colowick and they
established another long-lasting friendship. "Sidney was one of the most good-
natured persons I ever met. He had a critical mind and used it with the
greatest frankness but sometimes with a biting sarcasm that was made even
more impressive by his very soft voice." Once, when Gerhard was proudly
showing off his micro-Kjeldahl apparatus, used for nitrogen determinations,
he was praising the accuracy and the low blank values and other advantages.
Sidney listened quietly and at the end of this glowing description, said quietly,
"By the way, don't you think the connection between the condenser and the
distillation should really be tight?" He had noticed a leakage of vapor at
the junction.

Soon after his arrival, Gerhard could reciprocate by giving Sidney a useful
suggestion, based on his own experience, to use uranyl acetate for complete
precipitation of phosphate esters. He also gave Sidney some embarrassing
moments, as when, at the corner luncheonette, he complained noisily about
the plain white bread—instead of crisp rolls that would have been provided
in Germany—served with minute steak. On another occasion, after enduring
a long explanation about baseball, during the third game to which Sidney
had taken him, he said something indicating that he still thought that the
batter and pitcher were on the same team.

Gerhard enjoyed other associations in the relatively small Department of
Pharmacology. In addition to Carl and Gerty Cori and Sidney Colowick, the
staff included Associate Professor Helen Graham, who was also very active
in civic affairs and whose husband was a well-known pulmonary surgeon
(Evarts Graham), Assistant Professor Arnold Welch, and Mary Welch, who
worked with the Coris. Helping Gerhard to enjoy life in St. Louis, the Coris
introduced him to musicians with whom he could play in chamber groups.
During a session with one group, however, a member of the audience picked
up a newspaper and began to read it. Gerhard did not say anything, but the
disapproving look on his face was noted by the hosts and he was not invited
back to that group again.

A STIMULATING ENVIRONMENT FOR LEARNING

The scientifically exciting environment at Washington University was generated by outstanding faculty members of national and international prominence,

including some who were or would be Nobel Prize recipients. The Physiology Department was home to Joseph Erlanger, with whom Herbert Spencer Gasser (Professor of Pharmacology from 1921 to 1931, before his move to New York) had applied the cathode ray oscilloscope to the analysis of nerve potentials, and George Bishop, a pioneer in electroencephalography. Francis O. Schmidt, professor of zoology, developed x-ray diffraction methods for studies of macro-structure, and went on to head the Department of Biology at the Massachusetts Institute of Technology. Victor Hamburger was an outstanding embryologist who studied embryonal growth factors, and Michael Somogyi, director of the chemical laboratory of the Jewish Hospital, was a leading specialist in diabetes who made important observations on the actions of insulin. The dean of the medical school was Phillip Shaffer, a leader in study of metabolic disease who, together with Alexis Hartmann, developed a micro-method for glucose determination.

A most impressive learning experience was afforded by the Journal Club, conducted by Carl and Gerty Cori themselves, which met in the department's library for careful and critical examination of significant published papers. "It was, I can say, the spiritual center of the medical school." An event not to be missed except in case of an emergency, it was very well frequented, sometimes becoming so crowded that it had to be moved to the lecture hall. One had to be extremely well prepared and ready to participate in the discussion.

THREATENING NEWS FROM EUROPE

Meanwhile, with some of his family still in Germany, the situation in Europe had become ever more grave throughout 1938 and 1939.[9] The night of November 9, 1938, Kristallnacht, the night of broken glass, ended anyone's illusions about the possibility of Jews remaining in Germany. Nazi Storm Troopers destroyed 670 synagogues and 7,500 Jewish business properties, killed 92 people and arrested thousands of Jews, all in response to the action of a frenzied Polish youth, Hershel Grynzpan, who had killed a German-embassy official in Paris when his family was stranded between expulsion from Germany and denial of admission by Poland. Among the Jews arrested after Kristallnacht was Sepp Grünbaum, husband of Gerhard's sister Elizabeth. He was detained in Dachau for a few weeks and then released, perhaps because of illness or of his previous record as a district administrator and his service in the First World War. Soon after he was released, he and his family fled to England.

Europe was experiencing other turmoil as well. The Spanish Civil War, which came to an end with Franco's full victory by April 1939, had repercussions felt at Washington University in St. Louis, where a professor in the

---

[9] R. J. Evans, *The Third Reich in Power* (New York: Penguin Books, 2006).

Department of Microbiology was dismissed for having collected funds in support of the Spanish Republicans. Perhaps he was thought to be a Communist, but Gerhard noted that, especially in view of the grasp of power by Hitler and Mussolini in their nations, any "liberal" would have been sympathetic to the Republican cause in Spain, and many were so without being Communists. He was describing his own views.

Gerhard's mother and youngest sister Renate were still in Frankfurt. He had been saving money ever since his stipend at Queen's University exceeded his immediate needs, with the goal of getting them out of Germany. Renate did get to England during that year, via Kindertransport,[10] but his mother, Isabella, remained in Germany. At the beginning of August 1939, when the Coris went on vacation, Gerhard decided he would go to the Ozarks for a week. He spent a night in a pleasant small hotel in a picturesque village, but, worried about the situation, which was becoming worse day by day, he took the next bus back to St. Louis and tried to telephone his mother. He was horrified to learn there was no telephone connection to Germany, and he could not know what was happening. Two weeks later, he received a letter from England letting him know that she had managed to get a place on the last steamer that left from Germany to England, just three days before Germany's invasion of Poland. She was in Cambridge, England, with his sister Marion, who by then was married to Ernest Childs. Gerhard's mother and all his sisters and their spouses, and Edith Straus as well, were safe. He would focus efforts on bringing some of them to the United States, providing the necessary affidavits guaranteeing financial support.

Morris Cynkin, Gerhard's interviewer, questioned why Jews had waited so long to leave Germany. Speaking of his brother-in-law, Gerhard responded: "Don't forget, first of all, he was married and had two children who were 11 and 9 years old. As with most other Germans, his private fortune had completely disappeared in the inflation. He was high in the public service and dependent, for his future, on his pension and for the present, on his salary, his only income, which permitted a simple but comfortable living. In the world outside Germany there was a depression; it was very difficult to find jobs. He had diabetes, requiring continuous medical consultation. Also, he relied too much on the fact that he had been a combat officer—a captain—in the First World War; the army would protect him. He thought that eventually his children would have to go into exile, but by staying here he could help them more than by leaving Germany; it was really very difficult to take such a step for somebody who was not very wealthy." If one did leave, one was not allowed to take money or securities out of Germany, and was, in fact, required to pay

---

[10] For a description of Kindertransport, see: *United States Holocaust Memorial Museum Holocaust Encyclopedia*. Retrieved November 5 2012, from http://www.ushmm.org/wlc/en/article.php?ModuleId=10005260.

a steep "Reichsfluchtsteuer"—a "flight tax." When Gerhard's mother arrived in London, she had the equivalent of ten dollars; she had been forced to sell her property for next to nothing, and then was still asked to pay a "flight tax" based on a high valuation that said she still owned the property. Gerhard had been fortunate in not having a wife and children to support. By leaving Germany, he had been able to earn some money that would help his family members who were still there.

Gerhard also noted the mind-set of German Jews to stay in a set path in occupation and expectation. In contrast to that, what he found in America was remarkable flexibility with regard to earning one's livelihood *somehow*, even if it required a great change in the nature of one's work. He likened the Germans, and perhaps other Europeans, to domesticated cattle, being provided for by their system, and the Americans to wild buffalo, foraging where necessary to provide for self and family. In spite of difficult economic times, German Jews, in the period of recovery in the mid-1920s, felt they knew what their future and that of their children would be. They had a certain amount of capital that would make it possible for a son to study at the university and get a good job and would provide a dowry for a daughter. Above all, if they had a job—especially in government service, which included teaching in a high school or university—they would count on a pension. An academic department head had the contractual right to receive a pension, at age 65 or 70, with his full salary. At that time, pensions were not the usual expectation in America. Reluctant to give up the set pattern, some German Jews did not foresee, until quite late, the magnitude of the disaster coming at them.

Thirteen

On to Boston: The Odyssey Ends

SIEGFRIED THANNHAUSER

Gerhard prepared to move to Boston at the call of Professor Siegfried Thann-hauser,[1] also a German refugee. Thannhauser was an outstanding physician-scientist who carried on clinical work and, at the same time, headed a biochemical research laboratory at the Boston Dispensary, which had become part of the New England Medical Center, the major clinical affiliate of Tufts University School of Medicine. His offer provided the stable end-point for Gerhard's odyssey; Gerhard remained at the New England Medical Center and Tufts until his death in 1981. Siegfried Thannhauser was born to a prosperous Jewish family in Munich in 1885; his parents had a thriving business producing household ceramics. He showed talent in music and love for art and art history, but following completion of his Abitur at age 18, he opted for the study of medicine at the Ludwig-Maximilians-University School of Medicine in Munich. His major clinical teacher and mentor was Friedrich von Müller. Thannhauser became impressed with the contributions of chemis-try to the understanding of disease, though applications to clinical diagnosis were, as yet, not widespread. His appreciation of the importance of chemistry for medicine was reflected in his doctoral thesis, presented in 1910, on the subject of homogentisic acid, a metabolite of phenylalanine and tyrosine that accumulates in the disease alkaptonuria. Indeed, he next pursued formal study in chemistry, as the last student of the great Adolf von Baeyer, the head of the Department of Chemistry at Munich; his thesis, in 1912, was on the breakdown of hemoglobin.

Returning to Professor Müller's clinic, he continued chemical studies, focusing on gout and thus beginning his work on purines, nucleosides, and nucleotides, precursors of uric acid, which accumulate in tissues with this disease. He also studied diseased metabolism of glucose in diabetes, initiating a glucose-loading test for diabetes, and of cholesterol in liver disease. Still, he considered his clinical work and care of patients his main profession. At

---

[1] Biographical information and details of Dr. Thannhauser's medical and scientific contributions, along with appreciation by family members and colleagues, are provided by A. F. Hofmann and N. Zöllner, *Siegfried Thannhauser (1885–1962) Physician and Scientist in Turbulent Times* (Freiburg: Falk Foundation, 2004).

age 38, in 1924, he was appointed to head the outpatient department of the university in Heidelberg, and then was invited to the chair in Düsseldorf; and, in 1930, he was called to the chair of medicine in Freiburg where, a year later, he inaugurated the newly built University Clinic. There he began what became career-long interests and research on metabolism of lipids, particularly of sphingomyelin, and diseases associated with disturbances in their metabolism. By this time he was widely known, particularly for his *Textbook on Metabolism and Metabolic Diseases*,[2] published in 1929, a major work used by clinicians and by students studying clinical chemistry.

In 1932, Thannhauser recruited Hans Krebs to his department in Freiburg, where Krebs completed his classic work on the urea cycle by early 1933.[3] In mid-April of that year, while in the midst of studies on amino acid deamination, Krebs was dismissed from his post under the new Nazi law for "restoration of the civil service." In June, Krebs left Germany for Cambridge, England, where he had been offered a position by the great physiologist Frederick Gowland Hopkins. Thannhauser retained his faculty post at Freiburg a little longer than Krebs, perhaps because of his World War I service, for which he had been awarded the Iron Cross; but he too was dismissed in 1934, and could only conduct a private practice while weighing other possibilities. His non-Jewish wife, Franziska, read the signs of the brewing storm correctly and recognized the need to leave Germany. Early in 1935, the Thannhauser family emigrated to the United States. With support of the Rockefeller Foundation and the Emergency Committee in Aid of Displaced Foreign Physicians, he accepted a position offered by Dr. Joseph Pratt[4] at the Boston Dispensary.[5] There he established a research laboratory and served as associate clinical professor and then clinical professor of medicine of Tufts University School

---

[2] S. J. Thannhauser, *Lehrbuch des Stoffwechsels und der Stoffwechselkrankheiten* (München: J. F. Bergmann Verlag, 1929).

[3] F. L. Holmes, *Hans Krebs, the Formation of a Scientific Life: 1900–1933* (New York: Oxford University Press, 1991).

[4] Dr. Pratt, who helped place several refugees, received his MD from Johns Hopkins University in 1898, moved to Boston City Hospital, then spent a year studying with the internist Ludolf von Krehl in Germany. On return to City Hospital he studied typhoid fever, gout, function of the pancreas, and use of digitalis in heart disease. He moved to the Massachusetts General Hospital in 1903, joining the tuberculosis clinic. He introduced the use of group therapy in support of patient care. He became professor of medicine at Tufts University and chief of medicine at the Boston Dispensary, where he became interested in psychiatry and group psychotherapy, and in 1931 developed a program for rural medicine; H. H. Banks, *A Century of Excellence, The History of Tufts University School of Medicine 1893–1993* (Boston: Tufts University, 1993). The Pratt Diagnostic Clinic was founded at the Boston Dispensary in 1938, see *Journal of the American Medical Association*, 174(1960): 916–17.

[5] The Boston Dispensary, serving as a charitable outpatient service to provide for medical needs of the poor, was founded in 1796, the first hospital in Boston. Paul Revere was one of the original subscribers. See, W. R Lawrence. *A History of the Boston Dispensary* (Boston: John Wilson & Son, 1859). For a history of the Pratt Diagnostic Clinic of the Dispensary, see J. E. Garland, *An Experiment in Medicine; The First Twenty Years of the Pratt Clinic and the New England Center Hospital of Boston* (Cambridge, MA: Riverside Press, 1960).

of Medicine and senior physician at the New England Medical Center. In Boston, Siegfried Thannhauser and his wife, Franziska, with their three daughters, rebuilt their lives around their passion for art as well as around medicine and medical science.

At the Dispensary, Thannhauser met and helped to recruit other Jewish German refugees who were unable to hold their academic and hospital positions in Germany after the Nazi takeover. Two of the leading Dispensary physicians, Drs. Joseph Pratt and Samuel Proger, both of whom had trained in Germany, played important roles in establishing faculty positions for Thannhauser and others, gaining administrative support from Tufts University President John Cousens and especially his successor, President Leonard Carmichael, and financial support from the Rockefeller Foundation. In 1938, several members of the medical staff for the newly opening Pratt Diagnostic Hospital were identified as having fled Hitler's Germany.[6] Eventually there were at least thirteen, including Gerhard.[7]

At the Boston Dispensary, Thannhauser established a laboratory research program committed to studies of complex brain lipids and carried that interest into clinical work on lipidoses, diseases associated with abnormal accumulations of these materials. In focusing on brain lipids in Boston, he no longer followed up the important work his laboratory had done on nucleic acids in Germany, where he studied digestion of nucleic acids by human intestinal juice and intestinal bacteria. He had isolated the first crystallized deoxyribonucleotide—deoxyguanylic acid—and had found that *intestinal phososphomonoesterase* could be specifically inhibited by silver ions at very low concentration. In Europe he had also studied altered uric acid metabolism in patients with gout and disorders of carbohydrate metabolism in diabetes. Gerhard recalled that, with the relatively large amount of actual clinical work he had to do in his new setting, Thannhauser decided that it would be preferable, under the circumstances, to concentrate on complex lipids. These compounds included phospholipids based on glycerol (lecithin and cephalin and plasmalogens) or the nitrogen-containing backbone of sphingosine (sphin-

---

[6] H. Black, *Doctor and Teacher, Hospital Chief: Dr. Samuel Proger and the New England Medical Center* (Chester, CT: Globe Pequot Press, 1982); Joseph E. Garland, *An Experiment in Medicine. The First Twenty Years of the Pratt Clinic and the New England Center Hospital of Boston* (Cambridge, MA: The Riverside Press, 1960), 31–33; and Tufts University Archives, Medford, MA.

[7] The German physicians recruited to Tufts were: Heinrich Brugsch (medicine, rheumatology), Alice Ettinger (radiology), Joseph Fischmann (urology), Alfred Haputmann (neurology), Joseph Igersheimer (ophthalmology), Heinz Magendantz (medicine, cardiology), Martin Nothman (medicine, endocrinology), Bertha Ottenstein (biochemistry), Anna Reinauer (medicine), Jacob Schloss (medicine, gastroenterology), Gerhard Schmidt (biochemistry), Siegfried Thannhauser (medicine, biochemistry), and Richard Wagner (pediatrics). See B. D. Stollar, "A Way Out of Germany," *Tufts Magazine*, Winter 2014, 25. Available online at http://www.tufts.edu/alumni/magazine/winter2014/features/germany.html.

gomyelin)—and sugar-containing lipids based on sphingosine (cerebrosides).[8]

Gerhard arrived in Boston early in the spring of 1940, to a city still covered with snow piled so high that he "saw only the treetops of the Public Garden and the Boston Common." He found a room in a rooming house on Mt. Vernon Street on Beacon Hill, at five dollars a week, and reported to Thannhauser's laboratory. It had two good-size lab rooms, one balance room, a room for cleaning glassware, and one additional room where Dr. Joseph Pratt, director of the Pratt Diagnostic Clinic, carried on some work on intestinal enzymes.

As he began laboratory work in Boston, in April 1940, Gerhard was also aware that the war had come to the Western Front in Europe, with Germany crushing Belgium, the Netherlands, and France, occupying Denmark, and threatening England. He finally felt secure enough to take the steps required to bring his mother and youngest sister to America, and Edith Straus as well. He had written a letter to Edith, asking whether she would consider marrying him, a letter that Edith said "changed my life." She accepted his invitation. Gerhard provided affidavits of support for Isabella, Renate, and Edith, but immigration officials questioned whether his funds were sufficient, as he was also sending money to England for his sister Elizabeth, her husband, Sepp, and their two children, as a guarantee required by a family who had agreed to host them and in support of their education. Even with the savings he had retained, his salary was not large enough to support all these people. As a further complication, Edith's papers at the American Consulate in London became mixed up with those of another person with the same name. Eventually, these matters were resolved and arrangements were made for Isabella and Renate and, separately, for Edith, to sail to Boston in September.

FISHING EXPEDITIONS

Meanwhile, Gerhard lived alone in Boston. He worked weekends because he did not have a car and knew few people. He did, however, find a friendly scientific neighbor in the one other lab on the same floor of the building as his own: Nicholas Werthessen, who had been a student of Gregory Pincus at Harvard and was working in the laboratory of Charles Lawrence, a physiologist, at the Dispensary. Werthessen was an outgoing, witty, and energetic man who loved fishing and who determined to introduce Gerhard to that pastime. Gerhard accepted his invitation for a weekend of fishing off Cape Cod, though

---

[8] A question of particular interest for Thannhauser was whether sphingosine-based lipids may include a fatty acid ester linked to a hydroxyl group as well as the known amide linkage to the nitrogen of sphingosine, that is, whether there are more complex compounds of sphingomyelin, in which the second hydroxyl group of sphingosine, which was free in sphingomyelin as well as in cerebrosides—might be esterified. As sphingomyelin, lecithin, and cephalin are phospholipids, he also studied how they were modified by action of phosphatases.

he would have to provide his own gear. He went to the best sporting goods store in Boston, at the end of Washington Street near Scully Square, and bought an enormous deep-sea fishing rod for about $12, a lot of money for him in those days.

The next Saturday afternoon he and Werthessen drove to a place on the Cape where some people from Harvard's Biology Department had a summer home. The experience began most pleasantly, with a few drinks, dinner, and very stimulating conversation. "I enjoyed every minute of that. Then, when the sun was about to go down, Werthessen told me, 'Well, we have other things to do Schmidt,' and we had to go to our boat. I would have liked much more to stay that evening there and talk and sip drinks." Once in the rowboat, Werthessen handed Gerhard the oars. The rowing would be his task, a new experience for him, made especially difficult because Werthessen kept urging him to keep his fishing rod moving too—and to put new bait in place every so often. They started on a beautiful calm evening, but as the evening cooled and darkened, a breeze began to grow; clouds appeared and soon covered the whole sky. "A slight drizzle began and it was not quite as comfortable anymore. But Werthessen became more and more excited and every two minutes he shouted, 'There's a whole school—some stripers there—I think I see a whole school.' As far as I could see—I might have been wrong—these were only waves created by raindrops. . . . Occasionally I saw something happen to my rod because it got stuck in some weeds or something. No luck. About eleven o'clock or midnight, Werthessen concluded, 'We'll try another time. I think we'll go back to the car.' And I was very much relieved about that decision. I rowed all the time, and finally, with his direction, we landed at the right place; and what I liked even more was that he said, 'Let's go to one of the all-night hamburger places.' On the way, he said, 'I hope you had a nice time anyway.' I was so happy to be out of this drizzling rain and to go to this hot dog place. We had some refreshments and then we went back to the car, and while we were hardly more than ten steps from that joint, heading toward the car, there was in front of us a man with two big fish carried in both hands, and of course Werthessen got very excited and said, 'Where did you catch these?' He said, 'Well, not too far from the bridge.' I was looking forward very much to a relaxing drive home. I was tired out. Of course I couldn't drive; he drove the car. And so we had to go back, and of course now I had to go and row out there too—out of fairness. We tried to catch something until 7 o'clock in the morning." They caught no fish. It took a couple of days for Gerhard to recover from that outing. Not long after that experience, there came another venture, this time deep-sea fishing off Long Island, where Werthessen's father had a summer home, to which both he and Gerhard were invited. After a marvelous steak dinner, they boarded a large charter boat. Here Gerhard could use his deep-sea fishing rod. "And then we went out and at a certain place we stopped. And within 20 minutes I caught at least 35 mackerel. Finally it became boring." He did not take any home, because all he had in the Mt. Vernon apartment was a hotplate—no refrigerator. In retrospect, he enjoyed the fish-less experience on Cape Cod more than this one.

BRINGING MOTHER, SISTER, AND A FIANCÉE TO BOSTON

With Isabella and Renate coming to live with him in Boston, Gerhard's room on Mt. Vernon Street would not be large enough. He found a larger apartment in Brookline, one for the long term; it was still home for him and his family at the time of the interviews with Morris Cynkin 33 years later. He moved in about September 1st, 1940, having bought some furniture from Sears Roebuck and some second-hand furniture stores.

Isabella and Renate sailed from Liverpool to New York on the Cunard's HMS *Samaria*, part of a convoy that came under submarine attack during the voyage. Edith, on the other hand, sailed on another Cunard ship, the *Olympia*, from Liverpool to Quebec and Montreal, "where the German Jewish passengers were chaperoned on land by a group of B'nai B'rith members, feted with a luncheon, and brought to a train heading to Boston and New York. In Boston, Gerhard took me in his arms. He wore the same checked jacket he had worn in Gubbio."[9]

Isabella and Renate moved in with Gerhard, and Edith found housing in the South End, in the Ellis Memorial Settlement House. That was arranged by three very welcoming social workers, Misses Wightman, McCrady, and Tomkins. They gave Edith her first American Thanksgiving experience and invited her, Isabella, and Gerhard to their Christmas party in December. Knowing Isabella's musical skill, these women obtained a piano from one of their family members and shipped it to the Schmidt apartment in Brookline, at no cost to the Schmidts.

Gerhard and Edith planned to marry in January, but did not tell Isabella until just before the wedding. Edith did tell her friends at the Settlement House; they asked at which synagogue the wedding would take place and offered to arrange a rabbi's services at no cost to the couple. Gerhard and Edith, however, determined to be married simply and with minimal expense, by a justice of the peace. At wedding time, at the Settlement House on Chandler Street, Gerhard realized he was short one of the required two

---

[9]E. Schmidt, personal communication.

witnesses; the justice would not proceed. Gerhard rushed to the Boston Dispensary, not far away, and asked Dr. Martin Nothmann,[10] another of the German Jewish physicians at Tufts, to come with him for half an hour to serve as a witness; with Dr. Nothmann's presence, the wedding ceremony was completed, and Edith's friends provided a reception. Gerhard and Edith spent a few honeymoon days in Plymouth. They shared the next 40 years of their lives.

Edith moved into the Brookline apartment, joining Isabella and Renate as well as Gerhard. Isabella lived with them for nine years. Renate attended Brookline High School and then received a scholarship to Simmons College. She found accommodations in the home of a physician in Brookline in exchange for some household services. She pursued advanced studies in physics, married biochemist Arley Bever, and went on to a distinguished career with the National Aeronautics and Space Administration at the Goddard Flight Center in Maryland.

The Brookline household expanded with the birth of Michael in 1941, and Milton three years later. Gerhard and Edith set their roots in America firmly as Gerhard became a U.S. citizen August 1, 1944, and Edith followed suit March 19, 1945.

---

[10] Martin Nothmann, also a Jewish German refuges, was a good friend and prominent figure in the Department of Medicine. He had been an associate professor at the University of Breslau and then, from 1932–39 was physician-in-chief in the Department of Medicine at the Jewish Hospital in Leipzig. He was interned in the Buchenwald camp but released after a few months when invited to teach in Belgium. Soon after, he was brought to the Boston Dispensary in 1939 and resumed research in diabetes and, with Samuel Proger, in cardiology as a faculty member of Tufts University School of Medicine. Source: Tufts University Archives, Medford, MA.

Fourteen

Multiple Scientific Paths in Boston

INTESTINAL ALKALINE PHOSPHATASE

As the Schmidts settled into life in Brookline, Gerhard dove into research at the Dispensary. His first project, on *alkaline phosphatase*, was a good match for his developing interests as well as those of Thannhauser, who had published five articles the previous year concerning the increase of this enzyme activity in the serum of patients with certain diseases, Gerhard was fascinated by several aspects of *phosphatases*: their diversity, specificities, optimal conditions, and application as tools for studying structures of phosphate-containing substances, especially nucleic acids and nucleotides, but now phospholipids as well. The broader question of the roles of *phosphatases* in metabolism also stimulated him. It was easy to see that they were involved in breakdown of phosphate-containing organic compounds, just as other hydrolytic enzymes were involved in breakdown of polysaccharides, proteins, and lipids. A more difficult issue was how these complex structures were synthesized in living cells. At that time, very little was known about biosynthetic processes. Some scientists explored the possibility that synthesis could simply be a reversal of hydrolysis. At the Rockefeller Institute, for example, Bergmann had shown a sort of peptide bond synthesis with *papain;*[1] and in St. Louis Gerhard had shown reversibility of *phosphorylase*-catalyzed activity.[2] It was soon learned, however, that biosynthesis generally follows a more complex pathway than just the reverse of hydrolysis, with formation of intermediate compounds that reflected energy-dependent activation of substrates.

In view of Thannhauser's interests and his own, Gerhard began to study the *alkaline phosphatase* of intestinal mucosa. He set out to develop a large-scale purification to produce enough enzyme of high enough purity to allow careful study of its catalytic properties, including its activation or inhibition by various substances. He at last gave up the gravimetric method for measuring

---

[1] M. Bergmann and H. Frankel-Conrat, "The Role of Specificity in the Enzymatic Synthesis of Proteins. Synthesis With Intracellular Enzymes," *Journal of Biological Chemistry* 119(1937): 707. The subject was discussed later in M. Bergmann and J. S. Fruton, "The Significance of Coupled Reactions for the Enzymatic Hydrolysis and Synthesis of Proteins," *Annals of the New York Academy of Sciences* 45(1944): 409–23.

[2] C. F. Cori, G. Schmidt, and G. T. Cori, "The Synthesis of a Polysaccharide from Glucose-1-Phosphate in Muscle Extract," *Science* 89(1939): 464–65.

inorganic phosphate, the product of enzyme catalysis, and adopted a colorimetric method, using either the procedure developed by Fiske and Subbarow at Harvard University, or that of Delory in England.[3]

As a starting source of enzyme, he made a suspension of the mucosa scraped from 6 feet of calf intestine. He encountered a problem in trying to obtain soluble enzyme. The enzyme was bound firmly to cell walls, and it appeared in an insoluble lipid-rich layer lying over the water and water-soluble materials. He reflected that, according to the standards of the early 1940s, it was essential to prepare an enzyme in soluble form if one wanted to study it and hoped to publish a description of its properties. Looking back much later, he said, with a chuckle, "To submit a paper to an editor on an enzyme that had not been made soluble was about the same as to go on New Year's Eve to the Metropolitan Opera in a turtleneck sweater." Fortunately he found that, if he treated the suspension with *trypsin*, the *phosphatase* was released from cell surfaces; the enzyme, itself, was resistant to digestion by *trypsin*, so it became water soluble and retained its activity. With a 30% yield of total starting activity, he obtained a highly active, substantially purified enzyme.[4]

These data were presented in the *Journal of Biological Chemistry* in 1943, in the first article Gerhard published from his new base in Boston.[5] In addition to this work on intestinal *phosphatase* in his first years in Boston, he joined Thannhauser in writing a major review of research on the chemistry of lipins for the 1943 *Annual Review of Biochemistry*.[6]

He then had to determine his long-term scientific directions. Thannhauser had made clear that he could work on any project of his own choosing. That freedom was an opportunity to pursue his longstanding interests in nucleic acids and nucleotides, materials that Thannhauser had studied extensively in Germany. On the other hand, he felt a responsibility to consider that Thannhauser's main efforts in Boston were on phospholipids, particularly

---

[3]C. H. Fiske and Y. J. Subbarow, "The Colorimetric Determination of Phosphorus," *Journal of Biological Chemistry* 66(19925): 375–400; G. E. Delory, A Note on the Determination of Phosphate in the Presence of Interfering Substances," *Biochemistry Journal* 32(1938): 1161–62.

[4]Further purification steps included: removal of many impurities with insoluble alumina, to which phosphatase did not stick; fractional precipitation with ammonium sulfate; acetone precipitation; and dialysis. The enzyme was relatively nonspecific, cleaving substrates such as glycerophosphate, phosphopyruvate and all three phosphates of ATP. It was much less active in catalyzing hydrolysis of a diester, cleaving diphenylphosphate at less than one hundredth the rate at which the monoester phenylphosphate was split. Biological diesters, including DNA, and the phospholipids cephalin, lecithin, and sphingomyelin were not attacked at all. Gerhard was able to determine kinetic constants for the hydrolysis of the model substrate phenyl phosphate, and to show that magnesium had a slight stimulating activity, whereas phosphate, pyrophosphate, and cysteine were all inhibitory.

[5]G. Schmidt and S. J. Thannhauser, "Intestinal Phosphatase," *Journal of Biological Chemistry* 149(1943): 369–85.

[6]S. J. Thannhauser and G. Schmidt, "The Chemistry of the Lipins," *Annual Review of Biochemistry* 12(1943): 233–50.

focused on nervous system diseases and lipidoses (diseases of lipid metabolism), and that all of the funding for the laboratory was obtained by Thannhauser. He therefore would put at least part of his effort into supporting that work. He also recognized the important role of Dr. Samuel Proger[7] in supporting both his and Thannhauser's presence on the Dispensary staff, so he provided biochemical expertise for one of Proger's clinical studies. In 1945 he and colleagues published articles on each of these three themes: phospholipid degradation, a clinical study on anoxia with Dr. Proger, and nucleic acid chemistry.

PHOSPHOLIPID DEGRADATION

The 1945 phospholipid article in the *Journal of Biological Chemistry*[8] describing work done with assistance from Bessie Hershman, concerned the enzymatic breakdown of lecithin, a phospholipid, in mammalian tissues. After equivocal trials in which lecithin was added to extracts of various tissues, he found a very active lecithin breakdown activity in pancreas tissue extracts. This activity removed the fatty acid chains from the lecithin by hydrolysis. A major finding was the purification of the residual product, distinctive in being soluble in both water and ethanol, and proof of its identity as α-glycerophosphorylcholine (GPC). In GPC, nitrogen-containing choline is linked, via phosphate, to the hydroxyl on the third carbon of the three-carbon glycerol. The phosphate, forming esters with both the glycerol and the choline, is termed a phosphate diester. Reaction with sodium periodate revealed that the product was entirely the α-form that is, the diester was only on the third carbon of glycerol (see Figure 14.1).

Unlike the starting lecithin, which was totally resistant to hydrolysis by *alkaline phosphatase*, the diester GPC was cleaved by that *phosphatase* (though very much more slowly than was a phosphomonoester), to yield all three of its components separately: glycerol, choline, and inorganic phosphate. GPC was also hydrolyzed rapidly by acid (in HCl, 97°C), to glycerophosphate and free choline—a reflection of differing susceptibility of the two ester linkages

---

[7]Samuel Proger received his MD from Emory University in 1929 and came to the Boston Dispensary to train with Dr. Joseph Pratt. After a year in Boston, he pursued further training in Heidelberg and then returned to the Dispensary. An outstanding clinician, particularly in cardiology, he was chairman of the Department of Medicine at Tufts University from 1948 to 1971 and was active in the merger of several entities into the New England Medical Center; H. H. Banks, *A Century of Excellence. The History of Tufts University School of Medicine, 1893–1993* (Boston: Tufts University, 1993), 108. See also H. Black, *Doctor and Teacher, Hospital Chief: Dr. Samuel Proger and the New England Medical Center* (Chester, CT: Globe Pequot Press, 1982).

[8]G. Schmidt, B. Hershman, and S. J. Thannhauser, "The Isolation α- Glycerylphosphorylcholine from Incubated Beef Pancreas; Its Significance for the Intermediary Metabolism of Lecithin," *Journal of Biological Chemistry* 161(1945): 523–36.

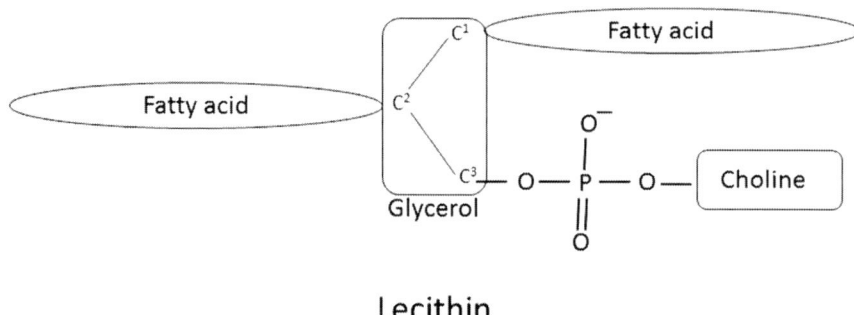

Lecithin

Figure 14.1. The phospholipid lecithin is composed of glycerol, with two fatty acids esterified at carbons 1 and 2, and a phosphorylcholine diester (this phosphate forms two ester links: to glycerol and to choline) at carbon 3. Removal of the fatty acids by hydrolysis yields the diester glycerylphosphorylcholine (GPC). Removal of the choline from GPC leaves the monoester glycerylphosphate (GP). Cephalin is an analogous structure; it has ethanolamine where lecithin has choline. It is broken down to glyceryl-phorphorylethanolamine (GPE) and then to GP.

of the diester; that is, the phosphate link to glycerol was stable while the link to choline was labile in acid. The lability of the bond between choline and phosphate depended on the diester structure, as phosphorylcholine monoester is quite stable in acid.

GPC was the first water-soluble organic phosphate product obtained from lecithin hydrolysis, and a novel feature in its purification was Gerhard's use of the recently introduced cation exchange resin Amberlite IR-100 to remove nitrogen-containing impurities. The final purification was particularly difficult because of its solubility; there were no conditions under which the GPC could be concentrated by precipitation or crystallized. Concentration by evaporation yielded an oily liquid. This inability to obtain crystals was frustrating. Gerhard was relieved, however, when, two to three years later, the outstanding chemist Erich Baer, at the Banting and Best Research Institute in Toronto, succeeded in synthesizing GPC—and his product too was a liquid that could not be crystallized.[9] Baer met Gerhard at the Federation Meeting that year and commended him on his isolation and identification of GPC as a breakdown product from lecithin.

As the GPC was formed in incubated suspensions of minced tissues at the same time that the amount of intact phospholipids declined, Gerhard and his colleagues concluded that GPC is a normal intermediate in the pathway of

---

[9]E. Baer and M.Kates, "L-α-glycerylphosphorylcholine," *Journal of the American Chemical Society* 70(1948): 1394–99.

lecithin breakdown, a residue that would exist after cleavage of two fatty acid chains from lecithin by the action of one or more *lipase* enzymes, distinct from *pancreatic lipase*. They wondered whether it may also be part of the pathway of synthesis.

A DIVERSION INTO CLINICAL RESEARCH

In research described in two articles published with Samuel Proger's laboratory,[10] Gerhard supported what might be considered an early exploration of translational medicine—transfer of new scientific findings into clinical applications—a current goal of 21st century research as well. Proger was interested in anoxia—the depletion of oxygen—and how it may be relieved. He was impressed with emerging biochemical insights on the importance of oxidative metabolism in maintaining ATP as a source of energy for both mechanical and metabolic work.[11] Knowing that the enzyme *cytochrome c* is important in oxidative metabolism, that it had been purified, was a relatively small protein, and was readily available, Proger proposed that injecting it intravenously into patients might counteract the ATP depletion of anoxia. Gerhard provided biochemical measurements showing that ATP levels in the heart and kidney of rats, measured by the amount of easily hydrolysable phosphate, were in fact decreased during anoxia, and less so in animals receiving an injection of 5 mg of *cytochrome c* before onset of anoxia. They also reported that the *cytochrome c* content of heart, liver, and kidney was increased in rats and rabbits after its intravenous injection, and under certain conditions, the arterial–venous difference in oxygen concentration was increased—in dogs but not in humans.

These findings provided hope for clinical application of *cytochrome c*. The hopes, however, were not sustained by studies in other laboratories. By 1948, Helmut Beinert and Kurt Reissmann, at the U. S. Air Force School of Aviation Medicine, cited several such attempts along with their own negative results.[12]

---

[10] S. Proger, D. Decaneas, and G. Schmidt, "The Effects of Anoxia and of Injected Cytochrome C on the Content of Easily Hydrolyzable Phosphorus in Rat Organs," *Journal of Biological Chemistry* 160(1945): 233–38; S. Proger, D. Dekaneas, and G. Schmidt, "Some Observations on the Effect of Injected Cytochrome C in Animals," *Journal of Clinical Investigation* 24(1945): 864–68.

[11] This concept had been expounded by Gerhard's friend Fritz Lipmann in a classic review article in 1941: F. Lipmann, "Metabolic Generation and Utilization of Phosphate Bond Energy," *Advances in Enzymology* 1(1941): 99. Lipmann, who had studied and worked with Otto Meyerhof in Germany, with P. A. Levene at the Rockefeller Institute in New York, and with Albert Fischer in Denmark, joined the Massachusetts General Hospital in 1941. See W. P. Jencks and R. V. Wolfenden, "Fritz Albert Lipmann. 12 June 1899–24 July 1986: Elected For Mem. R.S, 1962," *Biographical Memoirs of Fellows of the Royal Society* 46(2000): 333–44, 2000.

[12] H. Beinert and K. R. Reissmann, "Studies on the Incorporation of Injected Cytochrome C into Tissue Cells. I. Injection of Non-Radioactive Cytochrome C into Rats Previously Given Radioiron," *Journal Biological Chemistry* 181(1949): 367–77.

This outcome is not surprising, as the physiological role of *cytochrome c* depends on its being part of an intracellular structure, closely associated with other enzymes in the chain that eventually transfers electrons to oxygen. How would injected *cytochrome c* find its way to the true site of action, in amounts that could make a difference? It seems unlikely. Still, there has been a much more recent report, in 2007, that injection of *cytochrome c* protected cardiac muscle during sepsis, and the authors proposed that the injected enzyme did indeed make its way to the mitochondrial membrane of heart muscle cells.[13] An accompanying editorial cautions that there could be other mechanisms of action and the study requires verification.

This experience was at least the third time Gerhard had witnessed attempts to apply what were probably too-simple translations of science to the clinic, but it was the first in which he had been actively involved. The first occurred in Frankfurt, when Fischer-Wasel, the head of his Department of Pathology, tried to treat cancer patients with almost pure oxygen because he learned of Otto Warburg's discovery of increased anaerobic glycolysis in tumor cells. The second occasion was in Kingston, where Hendry Connell had given a preparation of bacterial proteolytic enzymes to patients with late-stage cancer. In each case, as in that of *cytochrome c*, there was nowhere near enough understanding of the underlying processes to allow a realistic evaluation of prospects for clinical efficacy.

THE SCHMIDT–THANNHAUSER METHOD FOR DNA, RNA, AND PHOSPHOPROTEIN

The third *Journal of Biological Chemistry* paper of 1945,[14] on the other hand, was a landmark article that became widely known and widely cited as a simple method for determining the amount of DNA, RNA, and phosphoprotein in a small sample of tissue extract. The idea was built on a foundation of Gerhard's long interest in the differing stability of phosphate ester bonds in nucleic acids and proteins, and was triggered by his encountering similar variations in phospholipids. He recalled that the then-known phosphate links in proteins, to the amino acids serine or threonine, were readily hydrolyzed by alkali, releasing inorganic phosphate into solution; he had used that property, originally shown by Plimmer in 1906,[15] in his earlier work with

[13] D. A., Piel, P. J. Gruber, G. J., Weinheimer, M. R. Courtois, C. M. Robertson, C. M. Coopersmith, S., Deutschman, and R. J. Levy, "Mitochondrial Resuscitation with Exogenous Cytochrome C in the Septic Heart," *Critical Care Medicine* 35(2007): 2120–27.

[14] G. Schmidt and S. J. Thannhauser, "A Method for the Determination of Desoxyribonucleic Acid, Ribonucleic Acid, and Phosphoproteins in Animal Tissues," *Journal of Biological Chemistry* 161(1945): 83–89.

[15] R. H. A. Plimmer and W. M. Bayliss, "The Separation of Phosphorus from Caseinogen by the Action of Enzymes and Alkali," *Journal of Physiology (London)* 33(1906): 439–61.

phosphoproteins in Kingston. He also knew that the diester bonds linking nucleotides in RNA were hydrolyzed by alkali, as first shown by Steudel and Peiser in Germany,[16] but in the degradation products the phosphate was still bound in the nucleotide structure; there was no release of inorganic phosphate. Further, he knew that DNA was not hydrolyzed at all by alkali, as shown much earlier by Kossel; it remained a large acid-insoluble polymer, whereas the same alkali conditions split RNA into individual nucleotides and released free inorganic phosphate from phosphoprotein.

Gerhard reasoned that, once pre-existing soluble phosphate compounds were removed in cold acid and phospholipids were removed by extraction in organic solvent, one should be able to measure the phosphate originating in phosphoprotein, RNA, or DNA by simply dissolving the residual tissue extract in 1N NaOH at 37°C for several hours (which would reduce RNA to mononucleotides and cleave phosphate from protein) and then adding acid to form a precipitate, separating acid-insoluble DNA from the acid-soluble nucleotides and inorganic phosphate. He determined: (1) total phosphorus in the starting alkali solution (ashing the sample), (2) total phosphorus in the acid-soluble fraction (ashing the sample), and (3) free inorganic phosphorus in the acid-soluble fraction (no ashing). From these values one can calculate: the phosphorus concentration of DNA [(1)-(2)]; phosphoprotein (3) and RNA [(2)-(3)]. He validated the method with assembled mixtures of purified RNA, DNA, and casein (a known phosphoprotein), and then applied it to tissue extracts to determine their RNA, DNA, and phosphoprotein content. When published data were available for comparison, this relatively simple method gave values similar to those obtained by more complex procedures involving extraction of nucleic acids from the tissues.

He gathered the required data over just a few months and submitted the manuscript on August 8; it was accepted immediately and published in the November 1 issue of the journal.[17] By coincidence, there was another paper on the determination of RNA and DNA in a tissue extract in the very same issue of the journal. It was from Walter Schneider of the McArdle Laboratory at the University of Wisconsin.[18] As in Gerhard's method, pre-existing acid-soluble and lipid substances were removed in Schneider's protocol, but the nucleic acids were then extracted from the residue with hot (90°C) trichloracetic acid, and DNA and RNA content were measured by color-generating

---

[16] H. Steudel and E. Peiser, "Uber die Hefe-nucleinsaure. III. Mitteilung," *Zeitschrift für physiologische Chemie* 120(1922): 292, cited in P. A. Levene, "Hydrolysis of Yeast Nucleic Acid with Dilute Alkali at Room Temperature. (Conditions of Steudel and Peiser)," *Journal of Biological Chemistry* 55(1923): 9–13.

[17] G. Schmidt and S. J. Thannhauser, "A Method for the Determination of Deoxyribonucleic Acid, Ribonucleic acid and Phosphoproteins in Animal Tissues," *Journal of Biological Chemistry* 161(1945): 83–89.

[18] W. C. Schneider, "Phosphorus Compounds in Animal Tissues: I. Extraction and Estimation of Desoxypentose nucleic acid and of Pentose Nucleic Acid," *Journal of Biological Chemistry* 161(1945): 293–303.

reactions: the deoxypentose-dependent DNA reaction with diphenylamine, and the ribose-dependent RNA reaction with orcinol. The Schmidt and Schneider methods were different, but their estimates of tissue nucleic acid content were similar. For different reasons, they both gave too-high values for brain. A striking finding was that the majority of nucleic acid in liver and pancreas, 80% or more, was RNA rather than DNA. Up until near that time, RNA was considered to be primarily a plant nucleic acid, and it was only in the late 1930s that Feulgen had demonstrated that plant cells contain both RNA and DNA, with RNA predominant in cytoplasm and DNA in nuclei. Gradually the presence of abundant RNA was becoming associated with cells having high levels of protein synthesis. Both methods were very useful in research in many laboratories, and both were highly cited over the years—nearly 4,000 times for Gerhard's paper.[19]

Gerhard recognized he could use the method to revisit his interest in sea urchin embryological development, something he had pursued in Naples when he left Germany and had tried to study even before then. With the new technique, he and a young colleague recently arrived from Germany, Liselotte Hecht, determined that, as Brachet had reported long before,[20] there is much more RNA than DNA in unfertilized sea urchin eggs. Gerhard also found that, in the first 24 hours of development after fertilization, the amount of RNA per embryo remained unchanged, whereas the amount of DNA per embryo increased 10- to 15-fold, reflecting the increase in cell number. By measuring these amounts per embryo he gave a picture that differed from Brachet's early report of a decrease in RNA during early development; Brachet had reported the RNA amount per unit weight of the embryo. This study clarified a discrepancy, at least in the interpretation of data, that was more than 10 years old.[21]

Continuing measurements of DNA in tissues, Gerhard obtained some lymphocyte samples from the outstanding hematologist at Tufts, William Dameshek. In this case it was possible to count the cells and he found the amount of DNA per cell was always the same. He did not have resources to extend the per-cell measurement to solid tissues, so he missed a very important generalization—that it would be the same in all somatic cells of an individual.

---

[19] ISI Web of Knowledge, accessed December 28, 2012, via Tuft's University library proxy to apps.webofknowledge.com.

[20] J. Brachet, "The Synthesis of Thymonucleic Acid During the Development of the Egg of a Sea Urchin," *Comptes Rendus des Séances de la Société de Biologie* 108(1931): 813–815; J. Brachet, "The Evolution of Pentose During the Development of the Egg of the Sea Urchin," *Comptes Rendus des Séances de la Société de Biologie* 108(1931): 1167–69. J. Brachet, *Embryologie Chimique* (Paris: Masson et Cie, 1944), 194 ff; G. Schmidt, "Über die Bindung der Purinbasen im unbefruchten Seeigelei," *Zeitschrift für physiologische Chemie* 223(1934): 81–85.

[21] G. Schmidt, L. Hecht, and S. J. Thannhauser, "The Behavior of the Nucleic Acids During the Early Development of the Sea Urchin Egg (Arbacia)," *Journal of General Physiology* 31(1948): 203–07.

He was gratified to learn, a few years later, that when Bovin did demonstrate a uniform DNA content per cell of all tissues in an animal, the measurements were made with Gerhard's method.

Gerhard's article was published with both him and Thannhauser as authors and the method has always been known as the Schmidt–Thannhauser technique, even though Gerhard noted that Thannhauser was not directly involved in its development. On the other hand, Gerhard reasoned that the idea for the RNA, DNA, and phosphoprotein method came to him while he was developing a procedure for distinguishing among phospholipids on the basis of alkaline lability, and he was doing that because it was in Thannhauser's field of interest.

In fact, however, nearly all Gerhard's papers published up to Thannhauser's death, in 1962, included Thannhauser as an author, regardless of how much or little the latter participated in the research. That was an understanding reached at the time Gerhard came to Boston. Grateful for a long-term position, and enjoying a warm relationship and friendship with Thannhauser, Gerhard did not resent this condition, even though it may have made it difficult for him to establish his own research identity. He recognized that Thannhauser obtained all the funding and was therefore, in a certain way, responsible for all the work carried out in the laboratory, and he was eager to contribute to the productivity required for regular renewal of that funding in Thannhauser's name. On the other hand, he realized that others responded differently to such a condition, recalling that Fritz Lipmann, who was at Cornell University from 1939 to 1941, in the Biochemistry Department headed by Vincent Duvigneaud, was invited by Thannhauser to come to the Boston Dispensary, but he would not accept Thannhauser's condition on authorship. For that reason, and probably others as well, Lipmann took a position as research associate in the Department of Surgery at the Massachusetts General Hospital in 1941.[22]

THE CURIOUS PHENOMENON OF METAPHOSPHATE FORMATION

A short time after the publication of the Schmidt–Thannhauser method, Gerhard received a stimulating visit by Winston H. Price, a brilliant young student from St. Louis. Gerty Cori had told Gerhard about him, and Gerhard was pleased to respond positively to a letter from Price asking for a meeting

---

[22] Lipmann then established his own biochemistry group at the hospital, became professor of biological chemistry at Harvard Medical School in 1949, and was awarded the Nobel Prize for Medicine or Physiology in 1953.

at the Dispensary during a visit to Boston Price was planning. At the meeting, Price impressed Gerhard with his detailed knowledge of scientific literature and clear thought. Among other things, Price brought Gerhard's attention to a new article written by Jean Brachet, with whom Gerhard had already had some differences in data interpretation. Brachet reported that when yeast cells were grown in a medium without phosphate and then transferred into a medium containing phosphate, there was a great increase in acidic polymer, detected by staining with a basic dye. Brachet concluded that the phosphate was incorporated into abundant new RNA synthesis, which would in fact be detected by the basic dye he used.[23]

Gerhard was particularly stimulated by the Price visit and his reference to the Brachet paper, enough so that he determined to verify the finding, an undertaking that certainly fit with his interest in nucleic acid metabolism. He and Liselotte Hecht obtained fresh yeast and verified the basic finding that transfer of yeast from phosphate-lacking to phosphate-containing medium was followed by uptake of phosphate into the cells into a polymeric form, insoluble in cold acid, which could be precipitated with barium sulfate. When he isolated and analyzed the polymer, however, he found it contained phosphate but no nitrogen—no purine or pyrimidine bases of RNA. In hot acid (HCl at 100°C), it was completely hydrolyzed to simple inorganic mono-phosphate. It was simply a polymer of phosphate, something known as meta-phosphate. It had been observed before in some fungal cells. He was anxious to publish his result and was encouraged to do so by Fritz Lipmann, who had heard that someone else was close to publishing the same finding. He and Hecht wrote a letter to the *Journal of Biological Chemistry* in October 1946. They continued to study the phenomenon and reported more details in a full article in the same journal three years later.[24]

They did not reach a conclusion on the physiological role of this synthesis, but suggested it may be involved in energy metabolism or phosphate group transfer. Metaphosphate formation was observed with several organisms in other labs, and, ten years later, Arthur Kornberg's lab studied the enzymology of the reaction in some detail.[25] Though it was an interesting phenomenon

---

[23] R. Jeener and J. Brachet, "Recherches sur l'acide ribonucléique des levures," *Enzymologia* 11(1944): 22.

[24] G. Schmidt, L. Hecht, and S. J. Thannhauser, "The Enzymatic Formation and the Accumulation of Large Amounts of a Metaphosphate in Bakers' Yeast Under Certain Conditions," *Journal of Biological Chemistry* 166(1946): 775–76; G. Schmidt, L. Hecht, and S. J. Thannhauser, "The Effect of Potassium Ions on the Absorption of Orthophosphate and the Formation of Metaphosphate by Bakers' Yeast," *Journal of Biological Chemistry* 178(1949): 733–42.

The authors quantified the uptake of phosphorus from the medium and showed that it was stimulated by potassium ions, with concurrent uptake of the potassium; and they tested the influence of several factors on the reaction.

[25] A. Kornberg, S. R. Kornberg, and E. S. Simms,. "Metaphosphate Synthesis by an Enzyme From *Escherichia Coli*," *Biochimica Biophysica Acta* 20(1956): 215–27.

that excited Gerhard's interest, polymeric metaphosphate did not find a place in schemes of physiological function. Gerhard summarized research on inorganic polyphosphates in a major review, which he presented at a Johns Hopkins symposium on phosphorus metabolism, and published in the proceedings of the symposium in 1951.[26] A few years later, in 1956, after studying the effect of phosphate and sulfate on protein and purine synthesis in yeast, Gerhard and colleagues reported that, in transfer from phosphate-deficient medium to one containing phosphate and sulfate, there was also, in fact, an increase in RNA purines, so Jeneer and Brachet had been partially correct in their conclusions; however, the RNA synthesis was overshadowed by the more abundant formation of metaphosphate.

RNA STRUCTURE STUDIED WITH RIBONUCLEASE AND PROSTATE ACID PHOSPHATASE

With the arrival of Ricardo Cubiles as a new research associate in the Thannhauser lab, Gerhard was able to return to nucleic acid structure studies. Thannhauser had accepted Cubiles, a recent medical graduate from Spain, for a research training position, assigned him to work with Gerhard, and suggested a project on cholesterol. Cubiles, however, was more interested in nucleic acids and joined Gerhard's work on application of *ribonuclease* and *phosphatase* in analyzing RNA structure. Gerhard's earlier work on RNA, for which he had to use intestinal *alkaline phosphatase*, was limited by the fact that the intestinal enzyme was optimally active at alkaline pH, at which internucleotide bonds in RNA are labile; in the enzyme solution breakdown products resulting from both alkali and *ribonuclease* activities were slowly degraded entirely to nucleosides and inorganic phosphate. He wanted a *monophosphatase* that could cleave only the terminal phosphate of oligonucleotide chains without destroying the RNA chains, under conditions in which both DNA and RNA are stable, permitting him to analyze which bases were present—and perhaps their sequence—in the chains. While still in Germany, he had observed a *phosphatase* active at neutral or acid pH, but did not have time to pursue the observation or publish about it then. He knew that another lab in Germany had studied an *acid phosphatase* obtained from human prostate gland[27] and he decided to prepare some of that enzyme in Boston.

---

[26] G. Schmidt, "Biochemistry of Inorganic Pyrophosphates and Metaphosphates," in *Proceedings of a Symposium on Phosphorus Metabolism, Volume I.*, eds. W. McElroy and B. Glass (Baltimore: Johns Hopkins University Press, 1951), 443–75.

[27] W. Kutscher and H. Wolbergs, "Prostataphosphatase," *Zeitschrift für physiologische Chemie* 236(1935): 237–40.

The problem was how to obtain fresh human prostate tissue. Prostatectomy was a frequent surgical procedure, but on removal, the organ was promptly dropped in formalin and sent to the pathology lab. Trying to recruit help at the hospital, Gerhard sat, bored, through weekly pathology conferences for a year. Dr. McMahon, the chief pathologist, asked surgeons to provide tissue for his research, but nothing came of the request. Real help came from another direction, from music. As he did wherever he was, Gerhard sought out collaborators with whom to play chamber music. He began meeting weekly in a trio, in which the violinist was the excellent urological surgeon Joseph Fischmann, also from Germany, who was affiliated with Tufts Medical School though he performed surgery at a different hospital.[28] Fischmann agreed to save and provide fresh tissue, which Gerhard could then obtain by going to that hospital. On some occasions, Fischmann brought the tissues, in a bag, to the trio's rehearsal at Gerhard's home. Gerhard would store it in the refrigerator and bring it to the lab in the morning. Together, they conducted a study on the amounts of *acid phosphatase* in prostate in normal and pathological conditions.[29]

Gerhard developed a method for purification of highly active, soluble, and relatively stable enzyme. As hoped, the enzyme, with a slightly acid pH optimum, pH 5.3, had almost no effect on intact RNA or other phosphodiester compounds and, as expected, it readily cleaved the phosphate of mononucleotides or the terminal phosphate of an oligonucleotide chain. This enzyme could, therefore, be used to determine the length of oligonucleotides, from either RNA or DNA, from the ratio of total phosphorus (the inorganic phosphorus measured after ashing the sample) to terminal phosphorus (the amount of phosphorus released by the *phosphatase* enzyme activity). The enzyme was also helpful in the preparation of oligonucleotides.

With combined action of freshly prepared crystalline *ribonuclease* and prostate *acid phosphatase*, Gerhard and Cubiles, along with several colleagues, confirmed the previous finding that *ribonuclease* action stops short of converting RNA entirely to mononucleotides, most of the products being short

---

[28] Joseph Fischmann was also a Jewish German refugee who found a position in the Tufts University School of Medicine. Born in Hungary, He earned his MD at the University of Berlin in 1926, trained in general surgery and urology, and served at St. Hedwig Hospital in Berlin from 1933 to 1936. He was able to leave Germany, became a research fellow at Harvard Medical School in 1939, joined the Boston Dispensary in 1940, and entered the Tufts faculty as instructor in urology in 1941. He eventually became associate clinical professor, the rank he held at the time of his retirement in 1959. From 1940 on, he was also on the staffs of the Mt. Auburn, Beth Israel, Boston State, and Massachusetts Women's hospitals; Tufts University Archives, Medford, MA.

[29] J. Fischmann, H. A. Chamberlin, R. Cubiles, and G. Schmidt, "Quantitative Determinations of Acid Phosphatase in the Prostate Under Various Normal and Pathological Conditions—Preliminary Report," *Journal of Urology* 59(1948): 1194–97.

oligonucleotides (chains of nucleotides linked by phosphodiesters). All mononucleotides that were produced were pyrimidine nucleotides. After *ribonuclease* digestion, *phosphatase* rapidly produced inorganic phosphate,
amounting to just about 25% of total phosphate, arising exclusively from
about one half of the pyrimidines. This study provided the first clear evidence
that *ribonuclease* shows a distinct selectivity, cleaving at pyrimidines, especially at pyrimidine–pyrimidine sequences, with a strong preference for one
of the two pyrimidines in the chain, either cytosine or uracil. The implications
of the results went beyond defining the enzyme selectivity. They showed that
RNA must consist of regions of consecutive purines and regions of consecutive
pyrimidines rather than any regular alternating purine–pyrimidine structure.

Gerhard presented his findings at the 1946 Federation meeting, the giant
annual combined meeting of several biological science associations in Atlantic
City. Strolling on the boardwalk, he met a leading chromatin biochemist,
Alfred Mirsky, of the Rockefeller Institute, who complimented him and invited
him to speak at the next Cold Spring Harbor Symposium on Quantitative
Biology, in 1947. Gerhard was excited by the prospect of being part of this
high-caliber annual meeting. After submitting a letter with some of the results
to the *Journal of Biological Chemistry*,[30] he prepared a more complete presentation for the Symposium and for publication in the proceedings.[31]

GROWING RECOGNITION IN SCIENTIFIC CIRCLES

He was not disappointed. As always, the 1947 Cold Spring Harbor Symposium
was most stimulating; and, with its theme of nucleic acids and nucleoproteins,
it was a superb forum for Gerhard's presentation. This scientific gathering
occurred just three years after Oswald Avery, Colin MacLeod, and Maclyn
McCarty had identified DNA as the physical carrier of genetic information.[32]
In the late 1940s, questions of how that information was coded in DNA
structure and then further organized in chromosomes were of supreme interest.[33] Participants were interested in his data on RNA structure and even

---

[30] G. Schmidt, R. Cubiles, B. H. Swartz, and S. J. Thannhauser, "The Action of Ribonucleinase on Yeast Nucleic Acid," *Journal of Biological Chemistry* 170(1947): 759–60.

[31] G. Schmidt, R. Cubiles, and S. J. Thannhauser, "The Action of Prostate Phosphatase on Yeast Nucleic Acid," *Cold Spring Harbor Symposia on Quantitative Biology* 12(1947): 161–67.

[32] O. T. Avery, C. M. MacLeod, and M. McCarty, "Studies on the Chemical Nature of the Substance Inducing Transformation of Pneumococcal Types: Induction of Transformation by a Desoxyribonucleic Acid Fraction Isolated from Pneumococcus Type III," *Journal of Experimental Medicine* 79, no. 2 (1944): 137–58.

[33] H. F. Judson, *The Eighth Day of Creation: Makers of the Revolution in Biology* (New York: Simon & Schuster, 1979).

asked him to give an informal talk beyond his programmed presentation. He also enjoyed the social life of the week-long meeting, which included a barn dance one evening and abundant opportunity for conversation.

Gerhard was excited to be in the midst of leaders in the field. Among many impressive people, he recalled especially John Masson Gulland, a chemist from England, who gave the symposium's introductory lecture and with whom he formed a personal friendship in those few days. From reading journal articles, Gerhard knew that Gulland had shown how a *phosphodiesterase* present in snake venom degraded RNA, yielding 5-nucleotides. He also knew that Gulland was a pioneer in using ultraviolet spectroscopy in studies of nucleic acids. From spectroscopic studies, Gulland had demonstrated that the ribose sugar in ribonucleosides is attached to the nitrogen in the 9 position of purines, overturning a previous conclusion, by P. A. Levene, that it was bound to the nitrogen at the 7 position; the difference is significant in assembling a structural model for nucleic acid. Most striking for Gerhard, Gulland described, in his lecture on DNA structure, some titration experiments, using spectrophotometry, giving evidence that the polymeric DNA structure involves more than the covalent linkages of nucleotides in a chain; Gulland actually named hydrogen bonds as an additional structural feature.[34] Five years later, Watson and Crick would propose a model structure for DNA[35] with the crucial insight into just how the hydrogen bonds were involved and, in 1960, Paul Doty, Julius Marmur, and colleagues, at Harvard, would give specific meaning —the separation of DNA strands (denaturation)[36] —to the spectroscopic titrations that Gulland had made.

Besides the molecular and cellular science at that Symposium, Gerhard had another revelatory experience. Arriving at Cold Spring Harbor, he saw, for the first time, horseshoe crabs, in great numbers on the beach. These were things he had known only from book illustrations. Seeing the actual creatures was a "Wiedersehen" moment, the kind of awakening experience he had known on just a few other occasions. It compared with "when you see that famous professor of biochemistry, whom you see only pictures, and finally you meet him and first shake hands with him." He decided he would have to bring some skeletons back to Brookline to show to his sons, Michael and Milton. He did not want to store decaying smelly skeletons through all of the week of the symposium, and decided he would just come back to the beach

---

[34] J. M. Gulland, "The Structures of Nucleic Acids," *Cold Spring Harbor Symposia on Quantitative Biology* 9(1947): 236–52.

[35] J. M. Watson and F Crick, "Molecular Structure of Nucleic Acids: A Structure for Deoxyribose Nucleic Acid," *Nature* 171(1953): 737–38.

[36] P. Doty, J. Marmur, J. Eigner, and C. Schildkraut, "Strand Separation and Specific Recombination in Deoxyribonucleic Acids: Physical Chemical Studies," *Proceedings of the National Academy of Sciences USA* 6(1960): 461–76.

and pick some up near the end of the meeting. Sadly, when he came back
early on the last day of the meeting, the horseshoe crabs had returned to the
ocean; there were none on the beach. He told his story of disappointment to
John Gulland before taking some time, during that last day of talks, to travel
into New York to visit with his good friend Sidney Colowick, who had taken
a position at the city's Public Health Research Institute. It was a prolonged
visit, leading Gerhard to take the last train of the night from New York City
back to Cold Spring Harbor. It reached that station at 1:30 a.m. With no cabs
in sight, "I had to walk from the station. It was a beautiful night—moonlight
night. The cottage where I stayed had no electricity—very nice cottages, but
somewhat 'roughing it.' I came into my dark room, only a beam of moonlight
came into the window, and on my bed I saw eerie figures. It looked like
something biological, and when I looked close it was an army of—a company
of—horseshoe crab shells all ordered carefully according to size, and after-
wards I learned that Gulland, after hearing my story, had collected them."
The boys got to see the horseshoe crabs.

Tragically, John Gulland was killed in a train accident in England on
October 26, just a few months after the symposium. Gerhard dedicated the
article he contributed to the symposium proceedings to Gulland's memory;
and the proceedings book opened with a memorial tribute. In another memorial
note in the journal *Nature*, Jesse P. Greenstein, then chief of the newly created
Laboratory of Biochemistry of the National Cancer Institute, agreed with
Gerhard's estimation, writing of Gulland: "He was easily the dominant figure
at that conference, and the charm and ease of his manner, his gentle critical
spirit, together with the scholarliness and incisiveness of his thinking, evoked
general admiration and affection."[37]

Recognition of Gerhard's work led to more exposure in the scientific world.
He was invited by Arthur Kornberg to present a seminar at the National
Institutes of Health in Bethesda, Maryland. Another public event came unexpectedly. The program for the annual national meeting of the American
Chemical Society, held in San Francisco in the spring of 1949, included a
symposium on nucleic acids, to be chaired by Columbia University's Professor
Erwin Chargaff, a leading nucleic acid chemist. Chargaff, however, learned
he would have to make an urgent trip to Europe at the time of the meeting
and sent Gerhard a telegram asking whether Gerhard would replace him as
chairman of the symposium. Gerhard was glad to accept. The speakers had
all been invited by Chargaff. "I just had to sit there as a chairman and see
that the discussion wouldn't last too long. And the program was very good."

---

[37] J. P. Greenstein, "Prof. J. Masson Gulland," *Nature* 161(1948): 87–88. See also R. D. Haworth, "John
Masson Gulland," *Obituary Notices of Fellows of the Royal Society* 6(1948): 67–82. Access online at http://
www.jstor.org/stable/768912.

In his recollection, however, the most impressive talk at the meeting was not in the session he chaired. It was, instead, a ten-minute presentation by Waldo Cohn, from the Oak Ridge National Laboratory, describing the separation of nucleotides by ion exchange chromatography; with this method, Cohn could even separate each nucleoside-3-phosphate from the corresponding nucleoside-2-phosphate.[38] This new technology greatly advanced the analysis of base composition and nucleic acid structure.

Gerhard knew and had befriended Cohn in Boston. Cohn, having studied at University of California, Berkeley, was at the Huntington Cancer Research Laboratory of the Massachusetts General Hospital and Harvard Medical School from 1939 to 1942. In fact, he and Gerhard shared a passion for music and played in chamber groups together. In 1943, Cohn moved to Oak Ridge, where he became senior chemist and group leader of the Biology Division of the National Laboratory. He also became a founder and first conductor of the Oak Ridge Symphony Orchestra. In the late 1940s Cohn invited Gerhard to accept a position at the Oak Ridge lab. It took Gerhard two years to decide definitively that he would stay in Boston at Tufts and decline that offer.

ASSOCIATION WITH TUFTS UNIVERSITY SCHOOL OF MEDICINE

By the late 1940s, Gerhard was well established at the Dispensary, but his relationship with Tufts University School of Medicine was just beginning. When he came to the Dispensary, the Medical School was located at 416 Huntington Avenue,[39] where today one finds the Northeastern University School of Law. "The buildings were not together; the Biochemistry department was in a garage-like annex. Physiology was a respectable building . . . I remember I would walk up wooden stairs; the floors were wooden floors—of course, very nice laboratories. The Department of Physiology was the only one in the medical school really interested in research at that time." That department was chaired by David Rapport, who, in the mid-1940s published several papers on alloxan-induced diabetes and thyroxine, with colleagues Attilio Canzanelli and Ruth Guild.

---

[38]The method was described in W. E. Cohn, "The Separation of Purine and Pyrimidine Bases and of Nucleotides by Ion Exchange," *Science* 109(1949): 377– 78; W. E. Cohn, "The Anion-Exchange Separation of Ribonucleotides," *Journal of the American Chemical Society* 1, 72(1950): 1471–78. For biographical information on Waldo Cohn, see "Human Radiation Studies; Remembering the Early Years," *Oral History of Biochemist Waldo E. Cohn, Ph.D.Conducted January 18, 1995* (Washington, DC: United States Department of Energy, Office of Human Radiation Experiments, June 1995). Retrieved November 14, 2012, from http://www.hss.energy.gov/ healthsafety/ohre/roadmap/histories/0464/0464toc.html#0464_Creating.

[39] H. H. Banks, *A Century of Excellence: The History of Tufts University School of Medicine* (Boston: Tufts University Press, 1993).

Frederick Reis, MD, a Tufts medical graduate, had served as chairman of the medical school's Department of Biochemistry since 1921. The department was very small—just Dr. Reis, Associate Professor Harry H. Powers, Assistant Professor Haroutioun H. Chakmakjian, and Instructor Elliott T. Adams—and the faculty carried out very little research, none of it in emerging fields of biochemical science. With no graduate program, its main role was teaching medical and dental students.

Gerhard, though appointed as assistant professor in biochemistry, did not have much contact with the department. Reis did invite him to present some lectures in the biochemistry course for medical students. Giving them an introduction to enzymes, "I talked about the discovery of coenzymes—Harden and Young's work—how they found that alcoholic fermentation needed a dialyzable cofactor, which had since been known as co-zymase. And then I asked them whether they understood what dialysis was. They said they hadn't heard about dialysis. So in my next lecture I prepared a demonstration with a collodion bag or a cellulose bag, and a saline solution on one side and a dialyzable Congo Red solution in the other side. Anyway, Reis must have heard of this demonstration—somebody must have told him, because when I came back I was told that Dr. Reis wanted to see me; and I was severely admonished that they show such experiments in the lab course, and one should not introduce any such thing in the lectures." Reis did not ask Gerhard to lecture again; there remained a distinctly cool relationship between them.

A marked change occurred in 1948. Reis had died and Halvor Christensen was recruited to be chairman of biochemistry.[40] Christensen, who had received his PhD from Harvard, had conducted research at Lederle Laboratories and as director of research at the Bassett Memorial Hospital in Cooperstown, NY, and returned to Harvard in 1946, was interested in nonprotein amino acids in body fluids and tissues and the relationship of free amino acid concentrations to tissue growth. At Tufts, and later at the University of Michigan and the University of California San Diego, he extended those interests to uptake of amino acids by cells. Soon after he came to Tufts he appointed Gerhard as research professor.

Christensen recruited some active scientists (Henry Sable, Eugene Knox), gave appointments to other scientists in the hospital (Peter Bernfield, Edward Frieden), and designed both research labs and student labs when the medical school moved, in 1949, to its downtown location at 136 Harrison Avenue, very close to the Boston Dispensary. He also began a graduate program. In 1950 his own student, Thomas Riggs, received the first PhD issued for investigative work carried out within the walls of the medical school.[41]

---

[40] Banks, *A Century of Excellence*, 79–80.
[41] T. R. Riggs, "Metabolic Acetylation and Pantothenic Acid Deficiency." PhD dissertation #548, Tufts University, 1950.

Christensen was succeeded as chairman of the Department of Biochemistry, in 1956, by Alton Meister, who greatly expanded research and the graduate program. He, in turn, was succeeded, in 1967, by Morris Friedkin.[42] Both Meister and Friedkin retained Gerhard's appointment in the department and involved him in both research and teaching, though his laboratory remained in the hospital.

MORE ON PHOSPHOLIPIDS, WITH GRADUATE STUDENTS

With an appointment in the new Department of Biochemistry and Nutrition, Gerhard resumed teaching activities and was able to accept graduate students. His first doctoral student, Lowell Greenbaum, however, was not from the biochemistry program. Greenbaum had met Gerhard at a symposium and, when he applied to and was accepted into the Department of Physiology program, he asked to do thesis research in Gerhard's lab; David Rapport, chairman of the Physiology Department, approved the request. A second student, Maurice Bessman, did come from the biochemistry program.

With Greenbaum, Gerhard resumed his work on α-glycerylphosphorylcholine, which he had identified as a product of acid hydrolysis of lecithin. Greenbaum developed a method for quantifying GPC in tissue extracts, based on measuring free choline (by its formation of a colored reineckate salt) before and after hydrolysis of aqueous tissue extracts. The difference represents the amount of choline originating from GPC. With this assay, he found high levels of GPC in fresh lamb liver and rapid formation of GPC in intestinal tissue

---

[42] Alton Meister, born in New York City in 1922, received his BS at Harvard (1942) and MD at Cornell (1945), headed a biochemistry laboratory at the National Institutes of Health until 1955, chaired the Department of Biochemistry at Tufts University School of Medicine from 1955 to 1967, and then chaired the Department of Biochemistry at Cornell University Medical College. He served as president of the American Society of Biological Chemists, was a member of the National Academy of Sciences, and was widely known for his research on the chemistry of amino acids and glutathione; N. Kresge, R. D. Simone, and R. L. Hill, "The Chemistry of Glutathione: The Work of Alton Meister," *Journal of Biological Chemistry* 282(2007): e30. See also H. H. Banks, *A Century of Excellence. The History of Tufts University School of Medicine, 1893–1993* (Boston: Tufts University, 1993), 99.

Morris E. Friedkin, born in Kansas City, received his BS and MS at Iowa State and his PhD in 1948 at the University of Chicago, as a student of Albert Lehninger. He was an NIH postdoctoral fellow in Copenhagen with Herman Kalckar 1948–49, a member of the Department of Pharmacology at Washington University in St. Louis, became chairman of the Department of Pharmacology at Tufts University School of Medicine in 1957, and of its Department of Biochemistry in 1967. He moved to the University of California, San Diego in 1968. He was elected to the National Academy of Sciences in 1978. He contributed extensively to research on folic acid metabolism and antileukemic therapy, the biochemical basis of the cell cycle and cell growth, the use of radioactively labeled molecules in biochemical research, molecular pharmacology, DNA synthesis, the structure and function of microtubules, and positron emission tomography. See Social Network and Archival Context, University of Virginia, http://socialarchive.iath.virginia.edu/xtf/view?docId=friedkin-morris-cr.xml; see also H. H. Banks, *A Century of Excellence*, 87.

that was allowed to incubate and undergo autolysis.[43] The results pointed to the importance of GPC as an intermediate in metabolic turnover of phospholipids. With graduate student Maurice Bessman, he studied the enzymatic cleavage of the phospholipid cephalin, yielding phosphoglycerolethanolamine (GPE) rather than GPC.[44]

Greenbaum also developed a method for measuring the total phosphoglycerol ester content of tissues, the sum of both GPC and GPE.[45] Both of these GP esters were, uniquely, soluble not only in acid but also with all other potential precipitants tested: copper sulfate, calcium hydroxide, and mercuric acetate. An advantage of this solubility is that all impurities can be removed by precipitation while GP esters remain in solution; a disadvantage is that one cannot concentrate them by precipitation. In application of the analytical method, impurities were precipitated from an acid-soluble extract of tissues, and then inorganic phosphate was measured in the remaining fluid with and without exposure to acid hydrolysis (to cleave choline or ethanolamine) and *prostatic phosphatase*. The difference in pre- and post-hydrolysis values corresponded uniquely to the phosphate derived from GP esters.

The GPE arose not only from cephalin, but also from a different group of glycerol-based compounds, present in brain tissue, which were then called "acetal phospholipids," also known as plasmalogens. Gerhard collaborated with Thannhauser and N. Boncoddo to develop an elaborate purification of "acetal phospholipids" and to characterize their structure.[46] Beginning each batch with ten pounds of beef brain, they obtained crystallized compounds, and found them to have a glycerol backbone with phosphoryl ethanolamine on the third carbon of the glycerol, and a formula corresponding to one long-chain alkyl group per molecule. This alkyl chain was rapidly cleaved from the structure by mercury, yielding an alkyl-aldehyde and GPE. The latter was crystallized and shown to be entirely α-GPE. When the aldehyde products were oxidized to fatty acids, both palmitic and stearic acid were formed.

---

[43] G. Schmidt, L. Hecht, P. Fallot, L. Greenbaum, and S. J. Thannhauser, "The Amounts of Glycerylphosphorylcholine in Some Mammalian Tissues," *Journal of Biological Chemistry* 197(1952): 601–09.

[44] G. Schmidt, M. J. Bessman, and S. J. Thannhauser, "The Hydrolysis of L-alpha-Glycerylphosphorylethanolamine, *Journal of Biological Chemistry* 203(1953): 849–53.

[45] G. Schmidt, L. M. Greenbaum, P. Fallot, A. C. Walker, and S. J. Thannhauser, "The Amounts of Glycerophosphoryl Esters in Some Tissues," *Journal of Biological Chemistry* 212(1955): 887–95.

[46] S. J. Thannhauser, N. F. Boncoddo, and G. Schmidt, "Studies of Acetal Phospholipides of Brain. III. The Fatty Aldehydes Present in Crystalline Acetal α-Phospholipide of Brain," *Journal of Biological Chemistry* 188(1951): 427–30; S. J. Thannhauser, N. F. Boncoddo, and G. Schmidt, "Studies of Acetal Phospholipides of Brain. II The α-structure of acetal phospholipide of brain," *Journal of Biological Chemistry* 188(1951): 423–26; S. J. Thannhauser, N. F. Boncoddo, and G. Schmidt, "Studies of Acetal Phospholipides of Brain. I. Procedure of Isolation of Crystallized Acetal Phospholipide from Brain," *Journal of Biological Chemistry* 188(1951): 417–21.

Prominent in this work was use of ion exchange resins to remove ions and neutralize preparations.

These brain lipids were called "acetal phospholipids" because their hydrolysis yielded, besides GPE, a long-chain aldehyde rather than a long-chain fatty acid (as in hydrolysis of a fatty acid ester), and acetals were known to yield aldehydes. Formation of an acetal, however, would require that only one long hydrocarbon chain be present per glycerol, linked to both the first and second carbons of a glycerophosphate molecule; that was the proposed structure at that time.[47] Although the elemental formula agreed with that proposal, it was learned later that, as in cephalin and lecithin, there were in fact two alkyl chains per molecule, one of them linked to the first carbon of glycerol as a vinyl ether, which also yields an aldehyde on hydrolysis, and the second as a standard carbonyl acid ester.[48]

With an abundant supply of GPE from the hydrolysis of "acetal phospholipids" Gerhard was able to test whether the lability of the choline-phosphate of GPC was more general, that is, whether the ethanolamine-phosphate bond of GPE would show the same property. Maurice Bessman, a graduate student in the Department of Biochemistry, joined in this work, which formed part of his thesis.

A large quantity of *prostatic phosphatase* had no effect on GPC or GPE. Neither did intestinal *alkaline phosphatase*, even though it did degrade yeast RNA. GPE was hydrolyzed by HCl (1 N, 100°C) even more rapidly than by NaOH (1 N, 100°C), releasing 99% of the ethanolamine and leaving glycerylphosphate; no inorganic P or free glycerol was formed. The product was a mixture of α- and β-glycerolphosphate (i.e., glycerol-3-phosphate and glycerol-2-phosphate), even though it had been shown that GPC was entirely of the α form.[49] There was an explanation for this puzzling result. At about that time Brown and Todd were studying the lability of RNA in alkali, attributing it to the formation of intermediate nucleoside, a 2-3- triester (which cannot form in DNA, which thereby is stable in alkali).[50] As they had pointed out, it was likely that same apparent migration occurred in glycerophosphate, similarly through formation of a cyclic (2,3) triester intermediate, which could then revert to either the 2- or the 3-linkage even though it all was α-glycerophosphate when first formed. This behavior gave a clue to why the P-choline or P-ethanolamine bonds are so labile within the glycerophosphate esters, whereas in free P-choline or P-ethanolamine they are stable.

---

[47] Thannhauser et al., "Studies of Acetal Phospholipids," 423–26.

[48] M. M. Rapport, "The Discovery of Plasmalogen Structure," *Journal of Lipid Research* 25(1984): 1522–27.

[49] G. Schmidt, M. J. Bessman, and S. J. Thannhauser, "The Hydrolysis of L-α -Glycerylphosphorylethanolamine," *Journal of Biological Chemistry* 203(1953): 849–53.

[50] D. M Brown and A. R. Todd, "Nucleotides. Part X . * Some Observations on the Structure and Chemical Behaviour of the Nucleic Acids," *Journal of the Chemical Society* (1952), 52–58.

Gerhard had also collaborated with Thannhauser in a study of lipids in Gaucher's disease, in which there occurs an accumulation of cerebrosides in cells, particularly in spleen and lymph nodes. Cerebrosides are phospholipids that also contain sugars—glucose and galacatose. An early proposal had been that cerebrosides were at abnormally high concentrations in blood and were deposited in tissues. Thannhauser, while he was still in Germany, had shown that blood levels were not, in fact, elevated; and he proposed the disease resulted from abnormal metabolism in the cells in which the lipids accumulated. Gerhard and he together confirmed that there was no elevation of serum levels and no accumulation in red blood cells. They developed a method for distinguishing between galactose and glucose in hydrolysates, using the ability of yeast to ferment glucose but not galactose. They demonstrated that both glucose- and galactose-cerebrosides accumulated in tissues, varying in proportion from patient to patient. They proposed that the disease was one of intracellular lipid metabolism, a disturbance in the balance between synthesis and breakdown of cerebrosides.[51]

As noted earlier, Gerhard had used varying lability of phosphate esters to great advantage in the method for measuring DNA, RNA, and nucleoprotein in one tissue sample simply by chemically measuring water-soluble and insoluble phosphate after incubation of a sample under varying conditions. Working with research associate Berta Ottenstein, graduate student W. A. Spencer, and several medical students, he applied the same principle in developing a method that could measure different classes of phospholipids in an emulsified tissue extract from sources such as brain, peripheral nerve, or heart muscle.[52]

Like Gerhard, Ottenstein was a Jewish refugee from Germany. She had studied both chemistry and dermatology and had been chief of the dermatology

---

[51] B. Ottenstein, G. Schmidt, and S. J. Thannhauser, 1948, "Studies Concerning the Pathogenesis of Gaucher's Disease," *Blood* 3(1948): 1250–58.

[52] G. Schmidt, B. Ottenstein, W. A. Spencer, K. Keck, R. Blietz, J. Papas, D. Porter, M. L. Levin, and S. J. Thannhauser, 1959. "The Partition of Tissue Phospholipids by Phosphorus Analysis," *AMA Journal of Diseases of Children* 97(1959): 691–708. The emulsified lipids are not water soluble. The extract is divided into three equal parts. Portion A is ashed and its total P content is measured. Portion B is shaken with N NaOH. In glycerophospholipids the carboxylic acid ester linkages are labile at high pH, so that this exposure to alkali releases water-soluble phosphate-containing compounds (GPC, GPE, GP-serine, GP-inositol). The plasmalogens remain insoluble, as the alkyvinylether is not cleaved by alkali. Portion C is incubated with mercuric acetate at pH 5.5. In plasmalogens, the vinylether linkage of one alkyl chain is highly susceptible to cleavage by mercury ions, releasing the alkyl chain. When this portion is subsequently treated with alkali, both the glycerol phospholipids and the residue of the plasmalogens release water-soluble phosphate, which is separated from the lipids with half-saturated ammonium sulfate at pH 5.5 and filtration. Thus, the water-soluble phosphate value for the Portion C is (glycerol lipid P + plasmalogen P). Portion B is glycerol lipid P alone, and the difference between B and C is the P from plasmalogens. The difference between Portion C and Portion A gives the amount of P from alkali-resistant phospholipids —that is, sphingolipids and ether phospholipids. See also G. Schmidt, B. Ottenstein, W. A. Spencer, C. Hackethal, and S. J. Thannhauser, "Quantitative Partition of Acetal Phospholipids and of Free Lipid Aldehydes," *Federal Proceedings* 16(1957): 832–35.

outpatient department (and the first woman to achieve a faculty position as lecturer) at the University of Freiburg. Siegfried Thannhauser headed its Department of Medicine and was aware of her work at that time. Dismissed in 1933, she fled first to Budapest, then to the University of Istanbul in 1936, where she became chief of the laboratory of the skin and cancer department. Thannhauser, in Boston, learned of her fears that Hitler would invade Turkey and reach Istanbul. Helped by the support of Tufts University president, Leonard Carmichael, she was recruited to the Boston Dispensary in 1944, as a research fellow.[53] She took up laboratory work with Gerhard.

APPLYING NEW ANALYTICAL METHODS

Throughout his career, through the decades of the mid-1930s to mid-1960s, Gerhard adapted enthusiastically to many new techniques available, especially to those that enhanced resolution, precision, and sensitivity of analytical methods. He and Morris Cynkin reflected on the long time it took, in many cases, from when principles and new methods were first discovered until they were widely accepted and applied. Examples of delays they discussed were use of ultraviolet light absorption by purines and pyrimidines for studies of nucleic acids, batch and column adsorption methods for purification of small and large molecules, paper chromatography, and preparation of high-molecular-weight DNA. Disruptions in communication during wartime certainly contributed to these delays. Considering the wide availability of scientific literature after those years, reasons for the delays were not clear; perhaps, in part, scientists held on to the familiar methods they learned. Certainly it took Gerhard a while to adopt colorimetric assays for phosphate in place of the gravimetric procedure he had learned from Embden. But he applied ion exchange and adsorption chromatographic methods as soon as he was aware of them, and took up paper chromatography and thin-layer chromatography quickly as well.

With new methods, he refined his earlier studies on RNA formation in sea urchin embryogenesis and addressed other questions, some of them off the main path of his research. With Maurice Liss, he applied sensitive chromatographic and spectrophotometric methods to confirm, with increased precision over previous analyses, that guanine was the major nitrogen-containing component in excrement of some spiders, and thus the major end product of their nitrogen metabolism.[54] With Ricardo Cubiles, he applied paper electrophoresis

---

[53] From correspondence in Tufts University Archives, Medford, MA.

[54] G. Schmidt, M. Liss, and S. J. Thannhauser, "Guanine, the Principal Nitrogenous Component of the Excrements of Certain Spiders," *Biochimica Biophysica Acta* 16(1955): 533–35.

and paper chromatography to update previously inconsistent analyses of the β-alanine-containing dipeptides carnosine and anserine, confirming their presence in skeletal muscle but absence in heart muscle of several species.[55] He, Maurice Bessman, and Mary Hickey used controlled centrifugation to separate granules of the yolk of chicken eggs from the clear soluble yolk plasma, and showed that all the phosphoproteins and 85% of the iron and calcium of yolk were present in the particles, whereas more than 70% of the phospholipids were in the soluble supernatant; most phospholipids were not closely associated with phosphoproteins.[56] He also took up the use of radioactive ^{32}P as a tracer for studying the uptake of phosphate from solution and its incorporation into phosphoproteins and phospholipids.[57] Some of these apparent diversions reflected excitement about the technology itself, reinforcing his long-standing interest in developing sound analytical methods.

MORE ON NUCLEIC ACIDS

Further work with Ricardo Cubiles that included research associate Krikor Seraydarian and graduate student Maria Seraydarian, refined the analysis of digestion of RNA by *ribonuclease*. They showed that 60% of all pyrimidines were released from RNA as mononucleotides by *ribonuclease*, and the rest of the pyrimidines were at the end of chains from which phosphate could be released by prostatic *monophosphatase*, that is, chains of purines with one pyrimidine at the end of each.[58] After removal of the accessible phosphate from these chains, the oligonucleotides reacted with periodate, indicating that the phosphate had been removed from the 2' or 3' position, leaving the pyrimidine attached to the chain through its 5'-phosphate. Thus the *ribonuclease* cleavage left a pyrimidine-3'-phosphate mononucleotide or a chain terminating with a pyrimidine-3'-phosphate. That is, the enzyme cleaved specifically to the bonds between pyrimidine–pyrimidine or pyrimidine–

[55] G. Schmidt and R. Cubiles, "Comparative Studies on the Occurrence of the Carnosine–Anserine Fraction in Skeletal Muscle and Heart," *Archives of Biochemistry and Biophysics* 58(1955): 227–31.

[56] M. J. Bessman, M. D. Hickey, G. Schmidt, and S. J. Thannhauser, "The Concentrations of Some Constituents of Egg Yolk in Its Soluble Phase," *Journal of Biological Chemistry* 223(1956): 1027–31.

[57] G. Schmidt and H. M. Davidson, "On the in Vitro Incorporation of ^{32}P-Phosphate into Phosphoproteins by Lactating Mammary Gland," *Biochimica Biophysica Acta* 19(1956): 116–20; G. Schmidt, L. H. Fingerman, H. M. Kreevoy, P. Demarco, and S. J. Thannhauser, "Incorporation of P32-Labeled Orthophosphate into Tissue Phospholipids of Intact Animals. Summary," *American Journal of Clinical Nutrition* 9(1961): 124–25.

[58] G. Schmidt, R. Cubiles, N. Zollner, L. Hecht, N. Strickler, K. Seraidarian, M. Seraidarian, and S. J. Thannhauser, "On the Action of Ribonuclease," *Journal of Biological Chemistry* 192(1951): 715–26.

purine sequences. The data also confirmed the previous conclusion that RNA structure was not an alternating purine–pyrimidine sequence, but rather some clusters of purines and clusters of pyrimidines of varying lengths.[59]

---

[59] G. Schmidt, R. Cubiles, and S. J. Thannhauser, "On the Nature of the Products Formed by the Action of Crystalline Ribonuclease (Kunitz's Ribonuclease) on Yeast Ribonucleic Acid," *Journal of Cell Physiology* (Suppl.) 38(1951): 61–70; G. Schmidt, R. Cubiles, N. Zollner, L. Hecht, N. Strickler, K. Seraidarian, M. Seraidarian, and S. J. Thannhauser, "On the Action of Ribonuclease," *Journal of Biological Chemistry* 192(1951): 715–26.

Fifteen

Mature Years, Honors, and Reflections

BECOMING DIRECTOR OF THE BIOCHEMISTRY LAB—AND A MISSED OPPORTUNITY

Siegfried Thannhauser reached the age of 70 in 1955; he had become Clinical Professor Emeritus in the mid-1950s. Officially, Gerhard became director of the Research Laboratories of the Boston Dispensary, but Thannhauser continued to work in the laboratory until his death in 1962. Speaking of that period, Gerhard recalled it as a time of incubating tension: "When he . . . retired in '53 or about that time and I was made director of the laboratory, he considered that completely . . . a sheer formality, that he was still the director of the lab. In a very disguised scientific manner, this came from my suggestion he just share [it] with me; . . . he wanted to show that . . . he was the boss. Our laboratory was fitted for experiments for him, and I had not . . . joined in and confirmed that."

The relationship between the two, formerly warm and friendly, cooled and became an uneasy one. Thannhauser continued to be an author on nearly all the manuscripts from the lab. Then a more troubling incident occurred. As Gerhard recalls, "This was in 1960. He (Thannhauser) was working for seven years on gangliosides, practically without any results. He . . . worked first with sphingomyelin, and then worked with acetal phospholipids—that was, by the way, on my suggestion; . . . and then he said now he will work on gangliosides after that." In 1959, Dr. H. Weicker, a physician from the University Hospital of Heidelberg, Germany, came to work with Thannhauser. "He was a very good man. The enormous contribution he made to the Thannhauser lab, and to Tufts in general, was that he brought . . . the equipment for thin-layer chromatography, which was completely unknown [here] at that time. . . . It had not come to this country.[1] [It] might be in the . . . literature but nobody knew it. Certainly it was completely unknown to Meister. Meister was completely paper chromatography. . . . And he [Weicker] said, 'Permit

---

[1] Thin-layer chromatography had been developed by Egon Stahl in the Dept of Pharmacognosy and Analytical Phytochemistry, Universität des Saarlandes, Saarbrücken, Federal Republic of Germany; E. Stahl, "Thin-Layer Chromatography," *Pharmazie* 11(1958): 633–37; E. Stahl, "Thin-Layer Chromatography. II: Standardization, Visualization, Documentation, and Application," *Chemische Zeitung* 82: 323–29.

me to put a small amount of your ganglioside preparation on this' . . . and
with the first chromatogram he found four spots of gangliosides. And of course
Thannhauser . . . was so excited . . . and he presented a paper on this in
1960 in Chicago." Thannhauser, however, because of his clinical work, was
particularly interested in Gaucher's disease, which involved cerebrosides
rather than gangliosides. Back in 1948, he and Gerhard had reported that
the cerebrosides accumulating in Gaucher's disease contained either only
glucose or a mixture of glucose and galactose, in contrast to those of brain,
which contain only galactose. Thannhauser used the new thin-layer method
to study spleen lipids from patients with Gaucher's disease and, with this
high-resolution method, he got an enormously complicated picture: there were
many spots on the chromatogram. "Anyway, he worked away on that with
many descriptive results and finally, . . . I said, 'Why not try to get cases of
Tay Sachs—where there's increase 10-fold, more, 40-fold . . . in the *ganglioside* level in brain?' And Thannhauser was absolutely opposed: 'No, we have
to concentrate on the spleen.'" That was because of the prominence of spleen
involvement in Gaucher's disease.

Gerhard took things into his own hands and wrote a letter. He knew people
at the Chronic Disease Hospital in Brooklyn[2] because he had given a paper
there, in a symposium, in 1959. Knowing that this specialized hospital had
collected Gauchers, Niemann-Pick, and Tay Sachs cases from all over the
United States, he wrote to Bruno Volk, a leader in studies of lipidoses, and
of Tay Sachs disease in particular, to ask whether he could get a sample of
brain from a Tay Sachs patient. In response, Volk sent Gerhard a whole brain,
from which Joel Dain, a research associate in the Schmidt lab between 1959
and 1962, prepared the gangliosides. "It was after Weicker left, in '61 when
I brought it up, and Thannhauser was so angry. After Weicker left, Dain was
there and he had to do it. He made a thin-layer chromatogram, and immediately
it showed there was a fifth spot, which was none of the other gangliosides,
except it was 40 times more intense than the others, and that was the Tay
Sachs ganglioside!—and Dain showed him that. . . . Anyway, Thannhauser
didn't pay any attention to that . . . I think it was mainly because I was behind
that experiment. He was so tense against me, because I was known probably
as the man that took over." Thannhauser had a very important result on this
chromatogram but did nothing with it.

When asked why people in the lab could not just go ahead and publish
and ignore Thannhauser, Gerhard responded, "Because we couldn't. He was
in that laboratory. At that time [he] had a technician, which he took away
from me. . . . Of course we couldn't. This was his work and this was his

---

[2]The Isaac Albert Research Institute, Kingsbrook Jewish Medical Center, Brooklyn, New York.

laboratory—and then he got the stroke and after three weeks he died. Under this condition, too late, I told Wagner [who had taken Weicker's place], now would he want to analyze this ganglioside from Tay Sachs, and got the analysis. [It] contained only one neuraminic acid, the correct analysis. Then I went one day to the library and in the [new issue of] *Research Communications* the whole thing was published by Svennerholm."[3] Gerhard had been scooped even though he had the result about a year earlier and could have received credit for it if he had published it when he first confirmed it.

Greatly disappointed at losing priority for this great discovery, Gerhard did try to publish his now confirmatory results, sending it as a preliminary communication to the *Journal of Biological Chemistry* and quoting Thannhauser's Chicago presentation demonstrating that the Boston lab had been working in this area for some time. He received a rejection letter from the editor, Morris Karnovsky, who commented that presentation at such a meeting is no evidence of priority. Gerhard did present the thin-layer chromatography analysis of gangliosides at a 1961 conference on sphingolipidoses in New York and in the published proceedings of the conference, with Dain, Weicker, and Thannhauser as coauthors.[4] That was the last of the articles to include both Schmidt and Thannhauser as authors.

THE LATER PHASE OF RESEARCH IN THE SCHMIDT LAB

After Thannhauser's death, Gerhard continued with research on both phospholipids and nucleoproteins, developing analytical techniques, particularly thin-layer chromatography, for separating and measuring tissue concentrations of particular phospholipids. During the 1960s he earned sustained research funding from the National Institutes of Health, the National Multiple Sclerosis Society, the National Science Foundation, The Charlton Fund, and the Godfrey Hyams Trust Fund. He was joined by several postdoctoral fellows and visiting scientists from Japan, establishing long-lasting connections, particularly with Dr. Teruji Tanaka, of the Jikei Medical School. Dr. Tanaka worked with Roy Keenan and Gerhard to develop a method, based on varying sensitivity of phospholipid components to hydrolysis, along with thin-layer chromatography,

---

[3] L. Svennerholm, "The Chemical Structure of Normal Human Brain and Tay-Sachs Gangliosides," *Biochemical and Biophysical Research Communications* 9(1962): 436–41.

[4] J. A. Dain, H. Weicker, G. Schmidt, and S. J. Thannhauser, "The Fractionation of Beef Brain Ganglioside into Several Components with Thin-Layer and Column Silica Gel Chromatography," in *Cerebral Sphinolipidoses: A Symposium on Tay-Sachs Disease and Allied Disorders*, eds. S. M. Aronson and B. W. Volk (New York: Academic Press, 1962), 288–99.

for measuring the marked variations in concentration of several distinct phospholipid fractions in brain and several other organs.[5] Kazuhiko Okabe also worked with Roy Keenan and Gerhard to show there are significant amounts of phytosphingosine (hydroxylated sphingosine, usually associated with plant phospholipids) in animal tissues.[6] Dr. T. Kitagawa joined in isolation of a new phosphoprotein, characterized by high phosphorus content, from the eggs of brown brook trout.[7]

Gerhard and colleagues turned again to cerebrosides, studying a mutant mouse strain (the "jimpy" mouse) with severe paralysis and a deficiency of myelinization in the central nervous system. They showed that while normal mice increased the amount of cerebrosides in brain myelin tenfold during development, the increase in jimpy mice was only twofold. There was no similar decrease in relative amounts of sphingomyelin or gangliosides.[8] The cerebrosides in the white matter of brain, constituents of myelin, are part of protein–lipid structures: proteolipids. The basic problem in the jimpy mouse turned out, still later, to be a genetically determined defect in the synthesis of the protein component, also called lipophilin.[9]

THE LAST WORKS: NUCLEOHISTONE AND NUCLEOPROTAMINES

Gerhard also extended and refined his research on action of nucleases on nucleohistone and nucleoprotamine. For doctoral-thesis research, graduate student Peter Cashion worked with visiting scientist Shigetaka Suzuki, John Joseph, Paul DeMarco, and another graduate student, Morris Cohen, on the digestion of nucleohistone DNA by *pancreatic DNAse*.[10] They identified a

[5] R. W. Keenan, G. Schmidt, and T. Tanaka, "Quantitative Determination of Phosphatidal Ethanolamine and Other Phosphatides in Various Tissues of the Rat," *Analytical Biochemistry* 23(1968): 555–66.

[6] K. Okabe, R. W. Keenan, and G. Schmidt, "Phytosphingosine Groups as Quantitatively Significant Components of the Sphingolipids of the Mucosa of the Small Intestines of Some Mammalian Species," *Biochemical and Biophysical Research Communications* 31(1968): 137–43.

[7] G. Schmidt, G. Bartsch, T. Kitagawa, K. Fujisawa, J. Knolle, J. Joseph, P. Demarco, M. Liss, and R. Haschemeyer, "Isolation of a Phosphoprotein of High Phosphorus Content from the Eggs of Brown Brook Trout," *Biochemical and Biophysical Research Communications* 18 (1965): 60–65.

[8] E. L., Hogan, K. C. Joseph, and G. Schmidt, "Composition of Cerebral Lipids in Murine Sudanophilic Leucodystrophy: The Jimpy Mutant," *Journal of Neurochemistry* 17(1970): 75–83.

[9] K. A., Nave, C. Lai, F. E. Bloom, and R. J. Milner, "Jimpy Mutant Mouse: A 74-Base Deletion in the MRNA for Myelin Proteolipid Protein and Evidence for a Primary Defect in RNA Splicing," *Proceedings of the National Academy of Sciences, U S A* 83(1986): 9264–68.

[10] G. Schmidt, P. J. Cashion, S. Suzuki, J. P. Joseph, P. Demarco, and M. B. Cohen, "The Action of Pancreas Deoxyribonuclease I (Deoxyribonucleate Oligonucleotidohydrolase, EC-Number 3.1.4.5.) on Calf Thymus Nucleohistone," *Archives of Biochemistry and Biophysics* 149 (1972): 513–27.

rapid phase of release of acid-soluble P, up to about 50% of the DNA content of nucleohistone, and then a very slow increase, along with formation of insoluble nucleoprotein in which there were equivalent amounts of basic (lysine and arginine groups in protein) and acidic groups (phosphates in DNA). They concluded that the rapidly digested component was DNA that was physically accessible to enzyme, corresponding also to DNA sites to which magnesium ions and low molecular weight polyamine compounds could bind; the number of magnesium binding sites decreased during the rapid phase of digestion. The enzyme-resistant nucleoprotein products showed only one peak in sucrose density centrifugation, corresponding to relatively small DNA–histone complexes, compatible with molecular weights in the neighborhood of 100,000 that had been suggested by other workers, and indicating that there is a uniformity of structure throughout the nucleoprotein, with DNA–histone segments of regular size separated by exposed DNA segments. These results corresponded well with the then-current research that was defining the nucleosome (a segment of about 145 DNA base pairs wound around a core containing a histone octamer) as a repeating subunit of chromatin structure.[11]

There was a disadvantage in working with *pancreatic DNAse* because it was relatively labile; it lost activity during the course of an experiment. Another *nuclease*, derived from micrococcal bacteria, did not have this problem. Nor did it require magnesium as a cofactor in the way the pancreatic enzyme did; a low concentration of calcium ions was sufficient. Morris Cohen, together with Paul DeMarco, took on a follow-up study on the action of *micrococcal nuclease* on both nucleohistone and nucleoprotamine.[12] The latter were more resistant to *nuclease* digestion than nucleohistone, and had fewer magnesium-binding sites. As with *pancreatic nuclease*, there was a rapid phase of DNA digestion of nucleohistone, releasing about 65% of the phosphorus into acid-soluble form; only 20% of nucleoprotamine DNA was released rapidly. With the high arginine content of protamines, there were almost equivalent numbers of basic and acidic charges in the starting nucleoprotein complex, but with much less accessible DNA. In other laboratories, the *micrococcal nuclease* was becoming the standard enzyme used for preparation of nucleosomes in research that dramatically increased understanding of chromatin organization during the 1970s.

---

[11] A. L. Olins and D. E. Olins, "Spheroid Chromatin Units (v Bodies)," *Science* 183, no. 4122 (1974): 330–32; R. D. Kornberg, "Chromatin Structure: A Repeating Unit of Histones and DNA," *Science* 184 (1974): 868–71.

[12] G. Schmidt, M. P. Cohen, and P. DeMarco, "The Action of Staphylococcal Nuclease (EC-number 3. 1. 4. 7.) on Thymus Nucleohistone (TNH) and on Some Nucleoprotamines," *Molecular and Cellular Biochemistry* 6 (1975): 185–94.

HONORS IN ACTIVE RETIREMENT YEARS

Shortly after Henry Mautner became chairman of the Department of Biochemistry and Pharmacology, Gerhard reached what was considered by the New England Medical Center Hospital to be time for retirement; he had to give up his lab, which was then on the fourth floor of the Rehabilitation/Emergency building of the hospital. He was not ready to leave research, however, and Mautner offered to accommodate him within the Department of Biochemistry, assigning him a laboratory in which he and his long-time assistant, Paul DeMarco, continued to investigate thin-layer chromatographic separations of tissue phospholipid components. It was at this time that Morris Cynkin invited Gerhard to talk about his life experiences and to record the conversations. They began October 4, 1971, meeting mainly in Cynkin's office in the Department of Biochemistry and continued for at least two years. Gerhard and DeMarco continued to work until Gerhard's final illness and death in 1981.

Gerhard had been elected to membership in the American Academy of Arts and Sciences in 1952; and, in recognition of his important contributions over many years, he was nominated for and elected to membership in the National Academy of Sciences of the United States in 1973. To recognize this milestone, Henry Mautner and George Brawerman organized a symposium and dinner in his honor, which took place March 5 that year, in the Cohen Auditorium on the Medford Campus of Tufts University. Participants and guests represented a large slice of the history of modern biochemistry. One session of the program was chaired by Carl Cori, and a second by Herman Kalckar. Speakers included Henry Mautner, Fritz Lipmann, David Nachmansohn, Erwin Chargaff, George Brawerman, Severo Ochoa—all friends of Gerhard—and Gerhard himself.[13] Cori, Lipmann, and Ochoa were all Nobel Prize awardees. Similarly distinguished guests were in the audience,[14] along with current faculty colleagues, former students and postdoctoral trainees, and current students. The session chairmen and speakers all had origins in Europe.

Introducing the program, Carl Cori looked back on important discoveries Gerhard had made while working with both Carl and Gerty Cori in St. Louis some 35 years earlier. He emphasized particularly Gerhard's purification of

---

[13] It is interesting that there were unplanned previous connections between Gerhard and Dr. George Brawerman when the latter came to Tufts with Henry Mautner in 1970. Brawerman, a Belgian Jew who had survived World War II as a child hidden in a youth camp, later studied in Brussels with Jean Brachet, some of whose experimental results and interpretations Gerhard had challenged but eventually rationalized. Brawerman then trained at Columbia University in New York with Erwin Chargaff, who shared Gerhard's interest in nucleic acids, contributing crucial chemical data used in the solution of DNA structure. As noted above, Chargaff knew Gerhard well and had invited him to serve as chairman of a scientific symposium; he was also a guest speaker at this symposium honoring Gerhard.

[14] Other guests included Konrad Bloch, Sidney Colowick, John Edsall, Eugene Kennedy, Boris Magasanik, Alton Meister, F. O. Schmitt, Jack Strominger, Bert Vallee, George Wald, snd Paul Zamecnik.

the enzyme *glycogen phosphorylase*, catalyst of an important step in the supply of blood glucose from glycogen stores in the liver. With purified enzyme, Gerhard discovered the reversibility in the *glycogen phosphorylase* reaction, and differences between the synthetic reactions with enzymes from skeletal muscle or liver. Cori also recalled the personal side of life in research, relating Gerhard's experience in setting out, by car, for Toronto to attend a Federation Meeting. Gerhard carried some enzyme with him so he could demonstrate the reaction it catalyzed—but called back en route to ask that someone bring some substrate, glucose-1-phosphate, which he had forgotten to take. Gerhard tested the enzyme activity every day up to the time of his presentation to be sure it remained active.

Fritz Lipmann, in his talk, looked back on the crucial insight of Embden in unraveling the role of sugar phosphates in the glycolysis pathway, an insight overlooked by members of the Meyerhof lab because they had a preformed idea that the sugar phosphates were by-products rather than true intermediates.

Gerhard's own talk was on the recent work on nucleoprotein structure, studied with pancreatic and micrococcal *nucleases*. At dinner and an informal reception at the Mautner home, friends and colleagues re-heard many Gerhard Schmidt "anecdotes" of the kind that had drawn Morris Cynkin to his project of interviewing Gerhard.

REFLECTIONS ON THE PAST

Coming to America

Gerhard had visited America in 1929, attending the International Congress of Physiology in Boston and adding a side trip to New York. He next passed through the United States on his way to Queen's University in Kingston, Ontario, in 1935. He then returned to work with P. A. Levene in New York in 1937, and stayed in the United States for the rest of his life. Having described the conditions he left behind in Germany, where in the mid-1930s he could no longer work or even participate in public life, attend movies, or even sit on park benches, he scattered remarks saying that: "this kind of ugliness you don't find in any other country. I never found it in Italy where there was also fascism, but certainly not anything like it would you find in the United States. There was Roosevelt, who was especially progressive, and Eleanor Roosevelt, and they were, for us, really idols at that time . . . certainly the ideas were for many refugees the symbol of liberalism . . . there was the Statue of Liberty, this is the country of freedom . . . it seemed really to be something totally different than the world of Prussia because you didn't hear anything or see anything of the army, nobody talked about the army; you hardly ever saw a soldier."

It came as a surprise to him when he heard that some Americans did not totally despise Hitler as he did. Some of his friends in New York might say, "What do you want with Hitler? He made his way up from a house painter to the Reich Chancellor and there must have been something good in him, otherwise he couldn't have gone so far. We learned that one thing that is characteristic for the United States, in contrast to the nobility-ridden countries in Europe, is that a man could come up from a very insignificant background and climb very high and that anything would be respected in the United States which could be achieved without great help from the outside." Later, it was even more of a surprise to see the development of a narrowing nationalism that could lead people to place bumper stickers saying "America—love it or leave it." Consistent with his liberal intellectual background, he was an Adlai Stevenson fan in the 1952 presidential election. He had no sympathy for the very conservative John Birch Society, formed in 1958.

Gerhard learned that vestiges of the old Germany remained in New York, particularly in Yorkville on the Upper East Side. "The last time I was in New York with Edith was '67, or might be '68. Anyway, Yorkville had one irresistible attraction for me. There was a pastry shop called Cafe Zeiger, with very genuine German, Viennese pastry . . . Baumkuchen, yeah, which is the ultimate German pastry. They also had specialties like Wienerschnitzel and good beer. And– yeah—one of the cafes on 86th St. had one type of German pastry called Schlagfinger—chimney sweets. Chimney sweets were tubes of very thin brittle dough about an inch in diameter and filled with whipped cream and covered with chocolate. And one of these cheaper cafes—they had their menus in the window—and then we got window shopping and we saw Schlagfingers—we saw these chimney sweets so I decided—we decided—it took some persuasion because Edith, in this respect, is a much stronger character than I. But finally she came with me, so we went to—the name of this pastry shop was Café Hindenburg. This was in the 60's—there was on every wall a big portrait of Hindenburg—on everything there was Hindenburg's face—the chief decoration everywhere—on every table was an ash tray with a golden base and an oval photograph of Hindenburg."

He remembered two things about academia that were new for him when he came to America. One was the very presence of private universities; virtually all the universities he knew in Germany were public institutions, mostly supported by the states or, in the case of Frankfurt, by the city. The second revelation was the different levels of prestige associated with certain American universities, particularly private universities like Harvard, in comparison with some of the state universities. In Germany the schools had particular strengths or favorable locations but all were generally viewed as being at a similar academic level.

The Good Years

In later years, Gerhard would look back on the period between the early 1940s and late 1950s as a particularly productive time in his career. He claimed a major factor in his productivity was that he did not have to be concerned with raising research funds. Thannhauser ensured that the laboratory received reliably renewed support from several sources: the Rockefeller foundation, the Godfrey Hyams Fund in Boston, the Charlton Fund administered by the Dispensary, and the Foundation for Infantile Paralysis (in the latter case, reliably up until 1951, when its interest turned to supporting polio vaccine development). Furthermore, shortly after World War II, the National Institutes of Health (NIH), and then the National Science Foundation, began to support—and even recruit—research in university laboratories. The Thannhauser lab received its first grant from the National Institute of Mental Health in 1948.

As the NIH began its grants program, it sent officials to various research laboratories to discuss emerging scientific questions with investigators and to encourage applications; and the research proposals were simple and short, explaining the nature of the problem to be studied and the general approach, without a mass of preliminary data or details of proposed experimental procedures. When one such emissary from the NIH, Dr. Irving Fuehr, visited Tufts and asked whether there were questions about the NIH grant that funded research in the lab, Gerhard asked him: "Yeah, I have one question, and it's kind of embarrassing, but sometimes you have the experience that you had a problem in which you put great hopes but it doesn't work out—you are coming to a dead end; you can't get rid of an impurity or something like that—and instead of spending a lot of time to settle that, you would like to start something else in the meantime." Fuehr said: "That's a very good question, and, as far as I can say, there's nothing written about it, but at least our attitude is the investigator knows best in which direction he should go." A year or two later, the application form, a bit more complicated, included the sentence: "If, during the work on the proposal, some leads turn up which are more promising than the original topic or object, the investigator should feel free to follow these leads." That was all; it was not necessary to get previous approval. The National Science Foundation had the same policy. "And this I think was very essential to the productivity of these years, because I know, for example, when I had found GPC as an enzymatic hydrolysis product of phospholipids in pancreas, and I wanted to work out a method to

determine it in tissues, and at that time for choline one of the best methods—
the most specific method—was the reineckate precipitation —reineckate, but
in a trichloroacetic acid filtrate —there were many substances which precipitate with reineckate and interfered with the method; and, when I didn't get
anywhere with this problem, which I started in '47 or so, I just dropped it
and did something else and took it up later when there was a way—
chromatography—to circumvent the problem."

By the mid-1950s Gerhard's work was recognized by invitations to contribute subject reviews and book chapters on both nucleic acids and phospholipids. He wrote on *nucleases* in the classic volumes, edited by Erwin Chargaff,
on nucleic acids,[15] and made many contributions on *phosphatase*, nucleic acid
preparation, and analysis and preparations of phospholipid components in
the classic volumes of *Methods in Enzymology*, edited by Sidney Colowick and
Nathan Kaplan.[16] He gave many research presentations, with corresponding
abstracts at meetings—mainly those of the Federation of American Societies
for Experimental Biology (FASEB), but also the American Chemical Society
and American Society for Clinical Investigation. In those years, his research
was moved forward by a team of graduate students, postdoctoral research
associates, medical students, and technicians.

WORKING WITH GERHARD

Gerhard's laboratory personality generated both reverence and apprehension
in his students and postdoctoral fellows. He worked for long hours, on both
weekdays and weekends, and expected similar intensity in his coworkers. He
also demanded the kind of respect that was paid to the professor in a German
laboratory, and could become very angry if that was lacking. Peter Cashion,
a student of Gerhard's in the late 1960s, was assigned to a project and
a room with Roy Keenan as his direct mentor. That room had the best
spectrophotometer in the department. One day Gerhard was using it for
periodic measurements of phosphate concentrations; he would come from
another room, make a measurement, disappear for some time and then reappear
to get another data point. Peter was also carrying on an experiment that
required a spectrophotometric measurement. Late in the day, he asked whether
he might be allowed to do one measurement in the time between those of the
Professor. Gerhard's response was not verbal. He pounded on the lab bench,
stomped his feet, and left the room, a sure sign of displeasure at a student

[15] G. Schmidt, "Nucleases and Enzymes Attacking Nucleic Acid Components," in *The Nucleic Acids*, ed. E. Chargaff (New York: Academic Press, 1955), 555–626.
[16] S. P. Colowick and N. O. Kaplan, eds., *Methods in Enzymology*, Vol. 3 (New York: Academic Press, 1957).

presumptuous enough to interrupt the Professor's work. Peter elicited a similar response by whistling at his work, something Gerhard did not tolerate in the lab.

On the other hand, Peter also experienced the warmth and charm of his professor. Gerhard invited him to come to New Hampshire, where Gerhard was on vacation—in a rustic lakeside cabin—for an overnight stay. There he saw the Professor completely relaxed, even without a tie, surrounded by nature, music, and books. He was impressed at the breadth and depth of Gerhard's culture and knowledge.

Gerhard was always to be addressed as "Dr. Schmidt," and wore a shirt and tie for all occasions. He carried this European sensibility into a hierarchical organization of his laboratory, in which he was clearly at the top. He often spoke just to his key laboratory assistant, John Joseph, who would then relay the words down to the students. At times, however, this pattern simply broke down and he would invite everyone in the lab to go for a beer at Jacob Wirth's, the nearby German restaurant, and would talk expansively about earlier days in the great biochemistry laboratories in both Germany and America. Or, at the end of a day of experiments on lipids of lobster nerves, he would gather everyone for dining on the rest of the lobster.

It was from John Joseph that Peter Cashion heard the solution to one of the lab's mysteries. Gerhard appeared in a neat summer jacket day after day. He confided to John Joseph, but not to the students directly: "I'll bet you think I'm wearing the same one every day. Actually I was at Filene's Basement and bought an armful of identical jackets. I just rotate them so I always wear a new-looking one."

Gerhard's unpredictable behavior was more than some associates could accept. Edward Masoro joined the lab in 1952 as a postdoctoral associate in biochemistry and lecturer in physiology, having been an assistant professor at Queen's University in Kingston, Ontario, for two years. He found Gerhard charming in social settings but a tyrant in the lab. After a few months, Masoro gave up the postdoctoral training and set up his own lab in the Department of Physiology at Tufts, where he stayed for the next ten years, before moving to head the Department of Physiology at the University of Texas Health Sciences Center in San Antonio.[17]

For those who persisted, however, Professor Schmidt remained a profound influence over many years—as well as a friend. Peter Cashion "treasured the experience" of having Gerhard as his mentor. The relationship was enhanced by the warmth shown them by Edith in the Schmidt home. Gerhard's students recognized his intellectual strength, breadth of knowledge, and love of music.

---

[17] From Science Careers, in *Science*. Accessed online November 29, 2012, at http://sciencecareers.science mag.org/career_magazine/previous_issues/articles/2003_01_08/noDOI.13920059766423123485.

With his cello and in social settings, he was a different person than the man they met intensely involved in the midst of his laboratory research.

Lowell Greenbaum, Gerhard's first student, recalled: "Gerhard had a temper which led him to stomp around the lab as we huddled together to find out what was wrong. It could be as simple as his not finding his lab coat in place. I was the unlucky one who had taken it by error. When I owned up to it, he remarked, "Ambitious aren't you!"

On the other hand, "At three o'clock every day, a large Erlenmeyer was set on a tripod and water boiled for instant coffee, which was poured into beakers. This was a great time for all in the lab because Gerhard (always addressed as Dr. Schmidt) would regale us with his experiences with Emden, and Gerty and Carl Cori. . . . He also told us about embarrassing situations that he got himself into." One situation he told them about was the swim from Kingston to Garden Island that had him showing up in a bathing suit at a faculty tea in progress.

"Schmidt was always available to review your results and to discuss future plans. One day, as we were reviewing my notebook on my project which involved the synthesis of glycerylphosphorylcholine, a precursor of lecithin, I remarked that yesterday's results were not good. Schmidt shot back a phrase which I have told all my own students; 'Young man, there are only results, not good or not bad!'"

"Gerhard loved his students despite our failings, which every graduate student goes through. He would invite to seminars, which the students gave, Nobel Prize winners like Fritz Lippman to listen for two boring hours to our research results. He was so proud of us. Some days he would insist we three sit down and he would fire biochemical and physiological questions at us."

Maria Seraydarian and Maurice Bessman joined the lab while Greenbaum was still there. Seraydarian recalled an episode that occurred when she was going to present a talk at the Federation Meetings in Atlantic City. The meetings were to start on Monday. She wanted to get there early, so she took off Saturday afternoon without telling Gerhard. Early Monday morning she was on the Boardwalk with Greenbaum and Bessman, who told her that Gerhard had been furious on learning she had taken off Saturday afternoon. They saw Gerhard approaching them on the Boardwalk. Greenbaum and Bessman peeled away, leaving Seraydarian to face what she anticipated would be an angry outburst. Instead, Gerhard was all smiles and charm, graciously asking her how she was enjoying herself in coming to the meeting.

A CAREER INTERRUPTED AND RESTARTED

Gerhard was a committed, driven scientist, who achieved much despite the interruption of his career just as he was reaching his potentially greatest years of scientific discovery. He was 32 years old, just two years into his hard-earned faculty position, when the Nazis came to power and he had to flee Germany. The next seven years proved to be a combination of good fortune—he was safe and able to work in laboratories of outstanding scientists—and, at the same time, limitations; he depended on short-term fellowships and, in Florence, Stockholm, New York, and St. Louis, he was directed to work on the topics of interest to the lab chiefs rather than on his own. He achieved published results in every one of these laboratories. Before reaching the Boston Dispensary and Tufts, it was only at Queen's University that he was an independent investigator, but that was for only two years, not enough time to build a career. Even when he reached the Dispensary, he felt obliged to develop projects fitting Thannhauser's interests in order to support funding for the Biochemistry Laboratory.

In most of his new starts, he had to adapt to new culture, climate, and language; beginning at Queen's, he wrote the rest of his scientific papers in English rather than German. It is difficult to imagine the enormity of the challenge he faced to be scientifically creative while harboring the lasting disappointment of his being expelled from his job, his country, and his German identity. He was not alone. These were challenges for a large number of refugees who had built scholarly and professional careers in pre-Nazi Europe and had been dismissed. Records of those helped by the Emergency Committee In Aid of Displaced Foreign Scholars noted how difficult this transition was for many of them, even while they were grateful for the help they received.[18] Thannhauser was one who succeeded in his new setting, but in a much smaller academic space than he had filled in Germany. In a letter in which he declined an invitation to return to his position at the University of Freiburg, Thannhauser spoke for many as he eloquently expressed this sense of loss and especially of dismay at being disowned by his German colleagues, who would avoid seeing him or even greeting him in the year before he left Germany.[19] Sharing these same feelings, Gerhard also succeeded remarkably in his scientific productivity, even in his short stays in several laboratories, and in his significant contributions once he was reestablished in Boston. He never returned to Germany.

AT HOME

Gerhard's son, Dr. Milton Schmidt, recalls a highly cultured home, filled with his father's love of music and great literature and his mother's love of art.

---

[18] S. Duggan and B. Drury, *The Rescue of Science and Learning*, 27–50.

[19] S. Thannhauser, Letter to Dr. Beringer, Dean of the Medical Faculty of the University of Freiburg, March 6, 1946. Tufts University Archives.

Musical evenings, with trios in which Gerhard was cellist, were prominent memories. Outstanding intellectuals in and beyond science were family friends. The cello and musical scores remain in Milton's home, long after his father's death. Gerhard's love of nature was also pervasive, especially evident during family vacations in New Hampshire. These very positive influences of family, music, nature, and literature were forces that could help Gerhard overcome his earlier sense of loss and disappointment.

Though they were not observant of religious practice, Gerhard and Edith retained the Jewish identity of their families. Both Michael and Milton observed the bar mitzvah tradition. At the ends of their lives, both Gerhard and Edith had Jewish funerals and burials.

Milton, a 1970 graduate of Tufts University School of Medicine, had the experience, perhaps a bit nerve-wracking at times, of attending his father's lectures. Milton practiced psychiatry in Massachusetts, and became director of the training program in psychiatry at the Newton Wellesley Hospital. Michael, a graduate of Tufts University in Medford, Massachusetts, and then City University of New York, became Clinical Supervisor of the Mental Health Clinic, Montefiore Medical Center, in New York.

With the children grown, Edith and Gerhard were able to simplify their living, moving in the late 1970s from a three-floor walkup to a smaller apartment in a building with an elevator, on Marion Street, still in Brookline. They were drawn to the new apartment building by the presence there of good friends, composer Herbert Fromm and his wife Leni, who were, like them, Jews who had fled Germany in the 1930s. Edith was volunteering in the print department of the Museum of Fine Arts. Gerhard continued to come into Boston to the laboratory. In his mid-70s he was to be found in the laboratory or the library, preparing a grant proposal for submission to the NIH. A diabetic, his health declined during these later years. In 1981, with family close by, his life ended quietly. His obituary in the *Boston Globe* of April 26, 1981, was headed "Gerhard Schmidt, 79, Biochemist; Hitler couldn't halt a noted career."

Edith continued to live in Brookline for some time, but then moved to La Jolla, California, to be with a friend. At age 89, she returned East, to live at the Scandinavian Living Center in Newton, Massachusetts. She continued to be a vital spirit, immersed in books, and newspapers and correspondence, keeping up with politics, delivering lectures on art history to other residents, and a regular winner at Scrabble. She enjoyed the love of two grandchildren, Alice and Eric, children of Milton and Homai. Before her 101st birthday, she attended the wedding of her granddaughter Alice, with a new outfit for the occasion, and led a toast to the couple. In the next year she began to take lessons in painting, achieving colorful pictures of flowers in watercolor.

On August 26, 2012, after a weekend visit by members of her family—and a game of Scrabble— Edith died, with family at her side, less than two months before her 102nd birthday.

AT REST, REMEMBERED

Beginning in May 1981, the Gerhard Schmidt Memorial Lectureship, sponsored by the Department of Biochemistry at Tufts University School of Medicine, has been an annual commemoration of Gerhard's life and work. The first invited speaker, Efraim Racker, was an outstanding biochemist, a member of the National Academy of Sciences, known for his research on the mitochondrial enzymes involved in ATP synthesis. Like Gerhard, he was a Jewish physician scientist, but he grew and studied in Vienna. He fled to Britain in 1938, soon after the Nazi takeover of Austria. The second year, Nobelist H. Gobind Khorana, a friend of Gerhard, presented the lecture. There has followed a distinguished series of speakers, including at least three more who would be honored with the Nobel Prize. The lectureship has been a highlight of the academic year at Tufts as well as a memorial to Gerhard Schmidt. It has also maintained connections with Gerhard's family—at least some combination of Edith, Michael, Milton, and Homai and their children were present at all of them up to the present—and with former trainees, particularly Dr. Teruji Tanaka, as well as with Tufts University faculty and students.

*****

Gerhard and Edith Schmidt both loved America and were deeply grateful that their adopted country allowed them to extend and so fully develop their lives. It also allowed them to retain their love of the music, literature, and culture of the Germany they once knew, including the writings of Johan Wolfgang von Goethe. A line from Goethe's *Wilhelm Meister's Travels* could apply to both Gerhard and Edith:

> „*Das Leben gehört den Lebenden an, und wer lebt,*
> *muss auf Wechsel gefasst sein.*"

"*Life belongs to the living, and he who lives must be prepared for change.*"

—Johann Wolfgang von Goethe, *Wilhelm Meister's Travels*

General Bibliography

Banks, Henry H. 1993. *A Century of Excellence, The History of Tufts University School of Medicine 1893–1993.* Boston: Tufts University.

Barraclough, Geoffrey. 1947. *Origins of Modern Germany.* Oxford: B. Blackwell.

Black, Herbert. 1982. *Doctor and Teacher, Hospital Chief: Dr. Samuel Proger and the New England Medical Center*, Chester, CT: Globe Pequot Press.

Brenner, Michael. 1996. *The Renaissance of Jewish Culture in Weimar Germany.* New Haven, CT: Yale University Press.

Childers, Thomas. 1983. *The Nazi Voter: The Social Foundations of Fascism in Germany 1919–1933.* Chapel Hill, NC: University of North Carolina Press.

Clare, George. 1982. *Last Waltz in Vienna: The Rise and Destruction of a Family: 1842–1942.* New York: Holt, Rinehart, and Winston.

Craig, Gordon A. 1982. *The Germans.* New York: Putnam.

Dawidowicz, Lucy. 1986. *War Against the Jews 1933–1945.* New York: Bantam Books.

Duggan, Stephen, and Betty Drury. 1948. *The Rescue of Science and Learning. The Story of the Emergency Committee in Aid of Displaced Foreign Scholars.* New York: MacMillan.

Elon, Amos. 1996. *Founder: A Portrait of the First Rothschild and His Time.* New York: Viking.

———. 2002. *The Pity of It All: A Portrait of the German-Jewish Epoch, 1743–1933.* New York: Metropolitan Books.

Evans, Richard J. 2004. *The Coming of the Third Reich.* New York: Penguin Books.

———. 2006. *The Third Reich in Power.* New York: Penguin Books.

Fest, Joachim. 2014. *Not I. Memoirs of a German Childhood.* Translated by Martin Chalmers. New York: Other Press.

Flaig, Ulrich. 1992. *Gustav Embden (1874–1933) und die Frankfurter physiologische Chemie.* Frankfurt am Main: Senkenbergischen Institut für Geschichte der Medizin der Johann Wolfgang Goethe-Universität.

Florkin, Marcel, and Elmer H. Stotz. 1975. *Comprehensive Biochemistry, Vol. 31: A History of Biochemistry Part III. History of the Identification of the Sources of Free Energy in Organisms*. Amsterdam: Elsevier.

Friedlander, Saul. 1997. *Nazi Germany and the Jews: Volume 1: The Years of Persecution 1933,–1939*. New York: HarperCollins.

Fruton, Joseph S. 1990. *Contrasts in Scientific Style: Research Groups in the Chemical and Biochemical Sciences*. Philadelphia: American Philosophical Society.

Garland, Joseph E. 1960. *An Experiment in Medicine. The First Twenty Years of the Pratt Clinic and the New England Center Hospital of Boston*. Cambridge, MA: Riverside Press.

Gay, Peter. 2001. *Weimar Culture: The Outsider as Insider*. New York: W. W. Norton.

Götz, Aly. 2014. *Why the Germans? Why the Jews?* English translation by Jefferson Chase. New York. Henry Holt.

Herzog, Dagmar. 1996. *Intimacy and Exclusion. Religious Politics in Pre-Revolutionary Baden*. Princeton, NJ: Princeton University Press.

Hofmann, Alan F., and Nepomuk Zöllner. 2004. *Siegfried Thannhauser (1885–1962): Physician and Scientist in Turbulent Times*. 2nd ed. Freiburg: Falk Foundation.

Holmes, Frederick L. 1991. *Hans Krebs, the Formation of a Scientific Life 1900–1933*. New York: Oxford University Press.

Larson, Erik, 2011. *In the Garden of Beasts: Love, Terror, and an American Family in Hitler's Berlin*. New York: Broadway Paperbacks.

Lipman, Fritz. 1971. *Wanderings of a Biochemist*. New York: Wiley.

Michaelis, Meir. 1978. *Mussolini and the Jews. German–Italian Relations and the Jewish Question in Italy 1922–1945*. Oxford: The Clarendon Press.

Nachma, Andreas, Julius H. Schoeps, and Hermann Simon, eds. 2002. *Jews in Berlin*. Berlin: Henschel Verlag.

Nachmanson, David. 1979. *German-Jewish Pioneers in Science, 1900–1933: Highlights in Atomic Physics, Chemistry, and Biochemistry*. Berlin: Springer-Verlag.

Peukert, Detlev J. K. 1987. *The Weimar Republic*. New York: Hill & Wang.

Reisman, Arnold. 2006. *Turkey's Modernization: Refugees from Nazism and Atatürk's Vision*. Washington, DC: New Academia Publishing.

Roth, Joseph. 1996. *What I Saw*. New York: W. W. Norton.

Sarfatti, Michele. 2006. *The Jews in Mussolini's Italy. From Equality to Persecution*. Madison, WI: University of Wisconsin Press.

Schorsch, Ismar. 1972. *Jewish Reactions to German Anti-Semitism, 1870–1914*. New York: Columbia University Press.

Shirer, William. 1960. *Rise and Fall of the Third Reich*. New York: Simon & Schuster.

Stern, Fritz. 1979. *Gold and Iron: Bismarck, Bleichröder, and the Building of the German Empire*. New York: Vintage Books.

——. 2006. *Five Germanys I Have Known*. New York: Farrar, Strauss & Giroux.

Taylor, Alan J. P. 1946. *The Course of German History*. New York: Coward-McCann.

Weitz, Eric D. 2007. *Weimar Germany: Promise and Tragedy*. Princeton, NJ: Princeton University Press.

Gerhard Schmidt Publications

1. Schmidt, G. 1928. "Über Kolloidchemische Veranderungen bei der Ermunding des Warmblutermuskels." *Arbeitsphysiologie* 1: 136–53.
2. Schmidt, G. 1928. "Über fermentative Desaminierung in Muskel." *Zeitschrift für physiologische Chemie* 179: 243–69.
3. Embden, G., and G. Schmidt. 1929. "Über Muskeladenylsäure und Hefeadenylsäure." *Zeitschrift für physiologische Chemie* 181: 130–39.
4. Schmidt, G. 1929. "Lactacidogen (Review)." In *The Enzymes*, edited by C. Oppenheimer, 1189. Berlin: George Thieme.
5. Embden, G., and G. Schmidt. 1930. " Über die Bedeutung der Adenylsäure für die Muskelfunktion: weitere Untersuchungen über die Herkunft des Muskelammoniaks." *Zeitschrift für physiologische Chemie* 186: 205–11.
6. Embden, G., and G. Schmidt. 1931. "Berichtigung." *Zeitschrift für physiologische Chemie* 197: 191–92.
7. Schmidt, G. 1931. "Über den Abbau des Guaninkerns durch die Fermente der Kaninchenleber." *Klinische Wochenschrift* 10: 165–67.
8. Schmidt, G. 1932. "Über den fermentativen Abbau der Guanylsäure in der Kaninchenleber." *Zeitschrift für physiologische Chemie* 185: 208–24.
9. Schmidt, G., and E. Engel. 1932. "Mikrobestimmungen von Purinsubstanzen in Geweben I. Mitteilung: Die Bestimmung des Guanins." *Zeitschrift für physiologische Chemie* 208: 225–36.
10. Schmidt, G. 1933. "Mikrobestimmungen von Purinsubstanzen in Geweben. 2. Mitteilung. Die Bestimmung des Guanins, des Adenins und der Oxypurine." *Zeitschrift für physiologische Chemie* 219: 191–206.
11. Rydh-Ehrensvärd, I., and G. Schmidt. 1934. "Über den Einfluss des Carotins auf den Guanasegehalt der Rattenmilz." *Zeitschrift für physiologische Chemie* 227: 177–80.
12. Schmidt, G. 1934. "Über die Bindung der Purinbasen im unbefruchteten Seeigelei." *Zeitschrift für physiologische Chemie* 223: 81–85.

13. Schmidt, G. 1934. "Zur Gewinnung der Dipeptidphosphorsäure aus Casein. Bemerkung zu einer Arbeit von P. A. Levene und D. W. Hill." *Zeitschrift für physiologische Chemie* 223: 86–88.

14. von Euler, H., and G. Schmidt. 1934. "Einfluss des Carotins (Vitamins A) auf den Puringehalt wachsender normaler und pathologischer Gewebe." *Zeitschrift für physiologische Chemie* 223: 215–28.

15. von Euler, H., and G. Schmidt. 1934. "Über Nucleoproteide der Fisch-Testikel." *Zeitschrift für physiologische Chemie* 225: 92–102.

16. Pentimalli, F., and G. Schmidt. 1935. "Über das Verhalten der Phosphor-fraktionen im Blutplasma sarkomkranker Hühner." *Biochemische Zeitschrift* 282: 62–73.

17. Schmidt, G. 1936. "A Chemical Difference Between Protein-Linked and Free Nucleic Acid." *Science* 83: 15.

18. Schmidt, G. 1937. "Effect of Nucleophosphatase upon Thymus Nucleohistone: Effect of Enzymes on Proteins with Prosthetic Groups I." *Enzymologia* 1:135–41.

19. Schmidt, G. 1937. "Growth-Stimulating Effect of Egg White and Its Importance for Embryonic Development." *Enzymologia* 4: 40–48.

20. Schmidt, G., and P. A. Levene. 1938. "The Effect of Nucleophosphatase on 'Native' and Depolymerized Thymonucleic Acid." *Science* 88: 172–73.

21. Schmidt, G., and P. A. Levene. 1938. "Ribonucleodepolymerase (The Jones-Dubos Enzyme)." *Journal of Biological Chemistry* 126: 423–34.

22. Cori, C. F., G. Schmidt, and G. T. Cori. 1939. "The Synthesis of a Polysaccharide from Glucose-1-Phosphate in Muscle Extract." *Science* 89: 464–65.

23. Cori, G. T., C. F. Cori, and G. Schmidt. 1939. "The Role of Glucose-1-Phosphate in the Formation of Blood Sugar and Synthesis of Glycogen in the Liver." *Journal of Biological Chemistry* 129: 629–39.

24. Schmidt, G., E. G. Pickels, and P. A. Levene. 1939. "Enzymatic Dephosphorylation of Desoxyribonucleic Acids of Various Degrees of Polymerization." *Journal of Biological Chemistry* 127: 251–60.

25. Schmidt, G., and S. J. Thannhauser. 1943. "Intestinal Phosphatase." *Journal of Biological Chemistry* 149: 369–85.

26. Thannhauser, S. J., and G. Schmidt. 1943. "The Chemistry of the Lipins." *Annual Review of Biochemistry* 12: 233–50.

27. Proger, S., D. Decaneas, and G. Schmidt. 1945. "The Effects of Anoxia and of Injected Cytochrome C on the Content of Easily Hydrolyzable Phosphorus in Rat Organs." *Journal of Biological Chemistry* 160: 233–38.

28. Proger, S., D. Dekaneas, and G. Schmidt. 1945. "Some Observations on the Effect of Injected Cytochrome C in Animals." *Journal of Clinical Investigation* 24: 864–68.

29. Schmidt, G., B. Hershman, and S. J. Thannhauser. 1945. "The Isolation of α-Glycerylphosphorylcholine from Incubated Beef Pancreas: Its Significance for the Intermediary Metabolism of Lecithin." *Journal of Biological Chemistry* 161: 523–36.

30. Schmidt, G., and S. J. Thannhauser. 1945. "A Method for the Determination of Desoxyribonucleic Acid, Ribonucleic Acid, and Phosphoproteins in Animal Tissues." *Journal of Biological Chemistry* 161: 83–89.

31. Schmidt, G., J. Benotti, B. Hershman, and S. J. Thannhauser. 1946. "A Micromethod for the Quantitative Partition of Phospholipide Mixtures into Monoaminophosphatides and Sphingomyelin." *Journal of Biological Chemistry* 166: 505–11.

32. Schmidt, G., L. Hecht, and S. J. Thannhauser. 1946. "The Enzymatic Formation and the Accumulation of Large Amounts of a Metaphosphate in Bakers' Yeast Under Certain Conditions." *Journal of Biological Chemistry* 166: 775–76.

33. Thannhauser, S. J., and G. Schmidt. 1946. "Lipins and Lipidoses." *Physiological Reviews* 26: 275–317.

34. Schmidt, G., R. Cubiles, B. H. Swartz, and S. J. Thannhauser. 1947. "The Action of Ribonucleinase on Yeast Nucleic Acid." *Journal of Biological Chemistry* 170: 759–60.

35. Schmidt, G., R. Cubiles, and S. J. Thannhauser. 1947. "The Action of Prostate Phosphatase on Yeast Nucleic Acid." *Cold Spring Harbor Symposia on Quantitative Biology* 12: 161–67.

36. Fischmann, J., H. A. Chamberlin, R. Cubiles, and G. Schmidt. 1948. "Quantitative Determinations of Acid Phosphatase in the Prostate Under Various Normal and Pathological Conditions—Preliminary Report." *Journal of Urology* 59: 1194–97.

37. Ottenstein, B., G. Schmidt, and S. J. Thannhauser. 1948. "Studies Concerning the Pathogenesis of Gaucher's Disease." *Blood* 3: 1250–58.

38. Schmidt, G., L. Hecht, and S. J. Thannhauser. 1948. "The Behavior of the Nucleic Acids During the Early Development of the Sea Urchin Egg (Arbacia)." *Journal of General Physiology* 31: 203–7.

39. Schmidt, G., L. Hecht, and S. J. Thannhauser. 1949. "The Effect of Potassium Ions on the Absorption of Orthophosphate and the Formation of Metaphosphate by Bakers' Yeast." *Journal of Biological Chemistry* 178: 733–42.

40. Schmidt, G. 1950. "Nucleic Acids, Purines, and Pyrimidines." *Annual Reviews of Biochemistry* 19: 149–86.

41. Schmidt, G. 1951. "Biochemistry of Inorganic Pyrophosphates and Metaphosphates." In *Proceedings of a Symposium on Phosphorus Metabolism*. Vol. I, edited by W. McElroy and B. Glass, 443–75. Baltimore: Johns Hopkins University Press.

42. Schmidt, G., R. Cubiles, and S. J. Thannhauser. 1951. "On the Nature of the Products Formed by the Action of Crystalline Ribonuclease (Kunitz's Ribonuclease) on Yeast Ribonucleic acid." *Journal of Cellular and Comparative Physiology* (Suppl.) 38: 61–70.

43. Schmidt, G., R. Cubiles, N. Zöllner, L. Hecht, N. Strickler, K. Seraidarian, M. Seraidarian, and S. J. Thannhauser. 1951. "On the Action of Ribonuclease."*Journal of Biological Chemistry* 192: 715–26.

44. Thannhauser, S. J., N. F. Boncoddo, and G. Schmidt. 1951. "Studies of Acetal Phospholipides of Brain. II. The α-Structure of Acetal Phospholipide of Brain." *Journal of Biological Chemistry* 188: 423–26.

45. Thannhauser, S. J., N. F. Boncoddo, and G. Schmidt. 1951. "Studies of Acetal Phospholipides of Brain. III. The Fatty Aldehydes Present in Crystalline Acetal α-Phospholipide of Brain." *Journal of Biological Chemistry* 188: 427–30.

46. Thannhauser, S. J., N. F. Boncoddo, and G. Schmidt. 1951. "Studies of Acetal Phospholipides of Brain. I. Procedure of Isolation of Crystallized Acetal Phospholipide from Brain." *Journal of Biological Chemistry* 188: 417–21.

47. Schmidt, G., L. Hecht, P. Fallot, L. Greenbaum, and S. J. Thannhauser. 1952. "The Amounts of Glycerylphosphorylcholine in Some Mammalian Tissues." *Journal of Biological Chemistry* 197: 601–09.

48. Schmidt, G., M. J. Bessman, and S. J. Thannhauser. 1953. "The Hydrolysis of L-α-Glycerylphosphorylethanolamine." *Journal of Biological Chemistry* 203: 849–53.

49. Schmidt, G. 1955. "[Thannhauser's Scientific Work, Since 1935]." *Deutsche Medizinische Wochenschrift* 80: 988–89.

50. Schmidt, G. 1955. "Acid Prostatic Phosphomonoesterase." In *Methods in Enzymology*, edited by S. P. Colowick, and N. O. Kaplan, 523–30. New York: Academic Press.

51. Schmidt, G. 1955. "Nucleases and Enzymes Attacking Nucleic Acid Components." In *The Nucleic Acids*, edited by E. Chargaff, 555–626. New York: Academic Press.

52. Schmidt, G., and R. Cubiles. 1955. "Comparative Studies on the Occurrence of the Carnosine–Anserine Fraction in Skeletal Muscle and Heart." *Archives of Biochemistry and Biophysics* 58: 227–31.

53. Schmidt, G., L. M. Greenbaum, P. Fallot, A. C. Walker, and S. J. Thannhauser. 1955. "The Amounts of Glycerophosphoryl Esters in Some Tissues." *Journal of Biological Chemistry* 212: 887–95.

54. Schmidt, G., M. Liss, and S. J. Thannhauser. 1955. "Guanine, the Principal Nitrogenous Component of the Excrements of Certain Spiders." *Biochimica et Biophysica Acta* 16: 533–35.

55. Thannhauser, S. J., J. Fellig, and G. Schmidt. 1955. "The Structure of Cerebroside Sulfuric Ester of Beef Brain." *Journal of Biological Chemistry* 215: 211–16.

56. Schmidt, G., M. J. Bessman, M. D. Hickey, and S. J. Thannhauser. 1956. "The Concentrations of Some Constituents of Egg Yolk in Its Soluble Phase." *Journal of Biological Chemistry* 223: 1027–31.

57. Schmidt, G., and H. M. Davidson. 1956. "On the in Vitro Incorporation of ^{32}P-Phosphate into Phosphoproteins by Lactating Mammary Gland." *Biochimica et Biophysica Acta* 19: 116–20.

58. Schmidt, G., K. Seraidarian, L. M. Greenbaum, M. D. Hickey, and S. J. Thannhauser. 1956. "The Effects of Certain Nutritional Conditions on the Formation of Purines and of Ribonucleic Acid in Baker's Yeast." *Biochimica et Biophysica Acta* 20: 135–49.

59. Schmidt, G. 1957. "Preparation of O-(l-α-Glycerylphosphoryl)choline, Phosphorylcholine, O-(l-α-Glycerylphosphoryl)-ethanolamine, and Phosphorylethanolamine." In *Methods in Enzymology*, edited by S. P. Colowick and N. O. Kaplan, 3: 346–58. New York: Academic Press.

60. Schmidt, G. 1957. "Preparation of Phosphopyruvic Acid." In *Methods in Enzymology*, edited by S. P. Colowick and N. O. Kaplan, 3: 223–28. New York: Academic Press.

61. Schmidt, G. 1957. "Determination of Nucleic Acids by Phosphorus Analysis." In *Methods in Enzymology*, edited by S. P. Colowick and N. O. Kaplan, 3: 671–79. New York: Academic Press.

62. Schmidt, G. 1957. "Preparation of Ribonucleic Acid from Yeast and Animal Tissues." In *Methods in Enzymology*, edited by S. P. Colowick and N. O. Kaplan, 3: 687–91. New York: Academic Press.

63. Schmidt, G. 1957. "Chemical and Enzymatic Methods for the Identification and Structural Elucidation of Nucleic Acids and Nucleotides." In *Methods in Enzymology*, edited by S. P. Colowick and N. O. Kaplan, 3: 747–75. New York: Academic Press.

64. Schmidt, G. 1957. "Colorimetric and Enzymatic Methods for the Determination of Some Purines and Pyrimidines." In *Methods in Enzymology*, edited by S. P. Colowick and N. O. Kaplan, 3: 775–81. New York: Academic Press.

65. Schmidt, G., M. J. Bessman, and S. J. Thannhauser. 1957. "Enzymic Hydrolysis of Cephalin in Rat Intestinal Mucosa." *Biochimica et Biophysica Acta* 23: 127–38.

66. Schmidt, G., B. Ottenstein, W. A. Spencer, C. Hackethal, and S. J. Thannhauser. 1957. "Quantitative Partition of Acetal Phospholipides and of Free Lipide Aldehydes." *Federation Proceedings* 16: 832–35.

67. Schmidt, G. 1959. "Nucleoproteins and Cancer." In *Physiopathology of Cancer*, 2nd ed., edited by F. Homburger, 707. New York: P. B. Hoeber.

68. Schmidt, G., B. Ottenstein, W. A. Spencer, K. Keck, R. Blietz, J. Papas, D. Porter, M. L. Levin, and S. J. Thannhauser. 1959. "The Partition of Tissue Phospholipides by Phosphorus Analysis." *AMA Journal of Diseases of Children* 97: 691–708.

69. Schmidt, G., L. H. Fingerman, H. M. Kreevoy, P. Demarco, and S. J. Thannhauser. 1961. "Incorporation of P32-Labeled Orthophosphate into Tissue Phospholipids of Intact Animals. Summary." *American Journal of Clinical Nutrition* 9: 124–25.

70. Dain, J. A., H. Weicker, G. Schmidt, and S. J. Thannhauser. 1962. "The Fractionation of Beef Brain Ganglioside into Several Components with Thin-Layer and Column Silica Gel Chromatography." In *Cerebral Sphinolipidoses: A Symposium on Tay-Sachs Disease and Allied Disorders*, edited by S. M. Aronson and B. W. Volk, 288–99. New York: Academic Press.

71. Schmidt, G., G. Bartsch, M. C. Laumont, T. Herman, and M. Liss. 1963. "Acid Phosphatase of Bakers' Yeast: An Enzyme of the External Cell Surface." *Biochemistry* 2: 126–31.

72. Schmidt, G., G. Bartsch, T. Kitagawa, K. Fujisawa, J. Knolle, J. Joseph, P. Demarco, M. Liss, and R. Haschemeyer. 1965. "Isolation of a Phosphoprotein of High Phosphorus Content from the Eggs of Brown Brook Trout." *Biochemical and Biophysical Research Communications* 18: 60–65.

73. Schmidt, G., E. L. Hogan, A. Kjeta-Fyda, T. Tanaka, J. Joseph, N. I. Feldman, R. A. Collins, and R. W. Keenan. 1966. "Determination of the Lipid Bases in the Lipids of Spinal Cord, Optic Nerve, and Sciatic Nerve of Some Species." In *Inborn Disorders of Sphingolipid Metabolism*, edited by S. M. Aronson and B. W. Volk, 325–59. Elmsford, NY: Pergamon.

74. Keenan, R. W., G. Schmidt, and T. Tanaka. 1968. "Quantitative Determination of Phosphatidal Ethanolamine and Other Phosphatides in Various Tissues of the Rat." *Analytical Biochemistry* 23: 555–566.

75. Okabe, K., R. W. Keenan, and G. Schmidt. 1968. "Phytosphingosine Groups as Quantitatively Significant Components of the Sphingolipids of the Mucosa of the Small Intestines of Some Mammalian Species." *Biochemical and Biophysical Research Communications* 31: 137–43.

76. Hogan, E. L., K. C. Joseph, and G. Schmidt. 1970. "Composition of Cerebral Lipids in Murine Sudanophilic Leucodystrophy: The Jimpy Mutant." *Journal of Neurochemistry* 17: 75–83.

77. Schmidt, G., P. J. Cashion, S. Suzuki, J. P. Joseph, P. Demarco, and M. B. Cohen. 1972. "The Action of Pancreas Deoxyribonuclease I

(Deoxyribonucleate Oligonucleotidohydrolase, EC-Number 3.1.4.5.) on Calf Thymus Nucleohistone." *Archives of Biochemistry and Biophysics* 149: 513–27.

78. Schmidt, G., M. P. Cohen, and P. DeMarco. 1975. "The Action of Staphylococcal Nuclease (EC-Number 3. 1. 4. 7.) on Thymus Nucleohistone (TNH) and on Some Nucleoprotamines." *Molecular and Cellular Biochemistry* 6: 185–94.

Index